The 1st and 2nd figures

of the complete group of 10 figures result from the manner of composition of the 2 half-words (prefix and suffix) as regards section(s) (A B C) and class as per following table:

First half-word (prefix)		and **last** half-word (suffix)		
		A same class	**B** same class	**C** same class
		First 2 figures	First 2 figures	First 2 figures
A	1st class	00	36	72
»	2nd »	01	37	73
»	3rd »	02	38	74
»	4th »	03	39	75
»	5 » »	04	40	76
»	6 » »	05	41	77
»	7 » »	06	42	78
«	8 » »	07	43	79
»	9 » »	08	44	80
»	10 » »	09	45	81
»	11 » »	10	46	82
»	12 » »	11	47	83
B	1st class	12	48	84
»	2nd »	13	49	85
»	3rd »	14	50	86
»	4th »	15	51	87
»	5 » »	16	52	88
»	6 » »	17	53	89
»	7 » »	18	54	90
»	8 » »	19	55	91
»	9 » »	20	56	92
»	10 » »	21	57	93
»	11 » »	22	58	94
»	12 » »	23	59	95
C	1st class	24	60	96
»	2nd »	25	61	97
»	3rd »	26	62	98
»	4th »	27	63	99
»	5 » »	28	64	
»	6 » »	29	65	
»	7 » »	30	66	
»	8 » »	31	67	
»	9 » »	32	68	
»	10 » »	33	69	
»	11 » »	34	70	
»	12 » »	35	71	

INDEX *to be used only for the **translation** of telegrams*

for half-words beginning with a **vowel**

if the 3rd letter of the half-word is a vowel, refer to ***Section C.***

a	e	i	o	u	
A or C 00	A or C 70	A or C 20	B or C 10	B or C 60	**b**
» 30	B or C 20	B or C 40	» 90	» 70	**c**
» 10	» 00	A or C 50	» 70	A or C 20	**d**
» 20	» 40	B or C 30	A or C 50	» 60	**f**
B or C 80	» 30	» 20	» 10	» 70	**g**
A or C 60	A or C 90		B or C 00		**j**
» 40	B or C 60	» 90	» 20	B or C 80	**k**
» 70	A or C 80	A or C 40	» 30	» 90	**l**
B or C 10	» 10	» 70	» 60	A or C 00	**m**
A or C 90	B or C 50	» 80	A or C 00	B or C 30	**n**
» 80	» 70	» 00	B or C 80	A or C 10	**p**
B or C 40	» 90	» 90	A or C 80	» 40	**r**
» 30	A or C 40	B or C 60	B or C 40	B or C 00	**s**
» 50	» 30	A or C 60	A or C 70	A or C 80	**t**
» 70	» 20	B or C 10	» 40	» 90	**v**
» 60		» 50	» 30	B or C 20	**w**
» 20	» 50	» 80	» 60	A or C 30	**x**
» 00	B er C 10	A or C 10	B or C 50	» 50	**z**

for half-words beginning with a **consonant**

if the 3rd letter of the half-word is a vowel, refer to ***Section C.***

b	c	d	f	g	h	j	k	l	m	n	p	r	s	t	v	w	x	z	
B C 50	B C 20	B C 60	B C 70	B C 40	B C 90	A C 00	B C 80	B C 10	A C 30	A C 40	A C 50	A C 60	A C 70	A C 80	B C 00	A C 90	A C 10	A C 20	**a**
B C 30	B C 00	D C 40	A C 70	A C 10	B C 70	B C 90	A C 00	A C 40	A C 20	A C 50	B C 50	A C 80	A C 60	B C 80	A C 30	B C 60	B C 10	A C 90	**e**
A C 70	B C 90	B C 20	A C 90	B C 60	A C 20	A C 40	A C 50	A C 30	B C 70	A C 10	B C 40	A C 00	B C 80	B C 10	B C 50	B C 30	B C 00	A C 60	**i**
A C 10	B C 80	B C 00	B C 10	B C 20	A C 30	A C 50	B C 50	A C 60	B C 40	B C 60	A C 90	B C 70	A C 40	A C 20	A C 00	B C 90	A C 80	B C 30	**o**
B C 80	A C 20	B C 90	A C 50	A C 90	A C 00	B C 70	A C 70	B C 20	B C 50	B C 40	A C 40	A C 30	B C 30	B C 00	A C 80	B C 10	A C 60	B C 60	**u**
A C 60	B C 40	A C 80	B C 00	A C 50		A C 20	B C 10	B C 50	A C 90	B C 20	B C 30	A C 70	A C 10	B C 70	B C 80	A C 00	A C 30	B C 90	**y**

Examples:

ce : page B 00
or page C 00

ig : page B 20
or page C 20

by : page A 60
or page C 60

VOLLERS'
10 Figure-System

compiled with the 2 letters minimum difference

Compiled and **published**
by
R. Vollers
Hamburg
10 Alsterdamm 10.

Price £ 5.-.- 2 copies, net cash, post free.

Printed by
C. L. KÜNCKE SÖHNE, HAMBURG.
1907.

VOLLERS'

10 Zahlen-System

10 000 Millionen Wörter von 10 Buchstaben, mit dem 2 Buchstaben Mindestunterschied ausgearbeitet u. genau den Beschlüssen des Londoner Telegraphen-Kongresses von 1903 entsprechend. Zahlengruppen von 0 000 000 000 bis 9 999 999 999.

Die Gesamtzahl besteht aus **12 Klassen** von je **900 Millionen** = **10 800 Millionen** Wörtern v. 10 Buchstaben; wie aus Seite 5 ersichtlich, ist sie auf **10 000** Millionen abgerundet.

Jede der 12 Klassen besteht aus 30 000 **Halb**wörtern von 5 Buchstaben (10 000 A, 10 000 B, 10 000 C) welche einander sowohl **vor-** wie auch **nach**gesetzt werden können und demnach 30 000 × 30 000 = 900 Millionen **ganze** Wörter von 10 Buchstaben ergeben.

In jeder Klasse haben die 30 000 Halbwörter den **2** Buchstaben Mindest-Unterschied und haben keine Morse-Ähnlichkeiten aufzuweisen wie z. B. zwischen **bn** —. . . | —. und **dr** —. . | .—. sowie auch **nf** —. | . .—.

Die **beiden Worthälften** eines **jeden** Wortes müssen **ein und derselben Klasse angehören,** damit der **2** Buchstaben Unterschied gewahrt bleibt, da **sämtliche** 360 000 **Halbwörter** dieses Buches sich selbstverständlich nur in mindestens **1** Buchstaben unterscheiden.

Die kompleten Wörter sind harmonisch, leicht aussprechbar u. daher ausgezeichnete Code-Wörter.

Folgende Buchstaben-Verbindungen kommen bei ihnen überhaupt nicht vor:

1) **jj, xx, yy, ly, yll,**
2) **aaa, bbb, ccc,** d. h. die Verdreifachung eines Buchstabens, ausgenommen „s."

Zu den 10-stelligen Zahlen liefert die 1te und 2te Worthälfte die **letzten 8 Zahlen** d. h. die 3te bis 10te Zahl und die Zusammensetzungsart der 2 Worthälften in Bezug auf Abteilung(en) (A, B, C) und Klasse die **1te und 2te Zahl.**

10 Figure-System

10 000 Millions of words of 10 letters compiled with the 2 letters minimum difference and in strict accordance with the decisions of the London Telegraph Conference of 1903. Figure groups from 0 000 000 000 to 9 999 999 999.

The total number is composed of **12 classes** of **900 millions** = **10 800 millions** of words of 10 letters; as will be noticed from page 5 they are reduced to a round total of **10 000** millions.

Each of the 12 classes consists of 30 000 **half**-words of 5 letters (10 000 A, 10 000 B, 10 000 C) which can be **prefixed** as well as **suffixed** to each other and which thus represent 30 000 × 30 000 = 900 millions of **complete** words of 10 letters.

The 30 000 halfwords of each class have the **2** letters minimum difference, and do not present any Morse similarities such as between **bn** —. . . | —. and **dr** —. . | .—. as well as **nf** —. | . .—.

The prefix and suffix of any word **must belong to the same class;** otherwise the difference of 2 letters would be reduced to 1 letter, there being, of course, only a minimum difference of **one** letter between **all of the half-words** viz: 360 000 existing in this book.

The complete words being harmonious and easily pronounceable are excellent code words.

The following combinations of letters do not occur in them:

1) **jj, xx, yy, ly, yll,**
2) **aaa, bbb, ccc, ddd** i. e. the triplication of any letter except **"s."**

The groups of 10 figures are composed as follows: The **last 8 figures** i. e. the 3rd to 10th figures result from the prefix & suffix and the **1st 2 figures** as per page 5 from the manner of composition of prefix and suffix with regard to section(s) (A, B, C) and class.

Die Nummerierung der 60 Seiten am Rande (A 00, A 10 etc.) ist sehr nützlich, denn die Randzahl stimmt jeweils mit der ersten horizontalen Zahl überein.

Gebrauchs-Anweisung.

Beim **Übersetzen** eines **Wortes** in eine **Zahl** sucht man zunächst mit Hülfe des Indexes und der Seitenangabe am Rande:
die **1te** Worthälfte mit der 3, 4, 5 u. 6ten Zahl und
» **2te** » mit der 7, 8, 9 u. 10ten Zahl
(die oberen horizontalen Silben sind nach ihren Vokalen geordnet)
und stellt sodann auf Grund des Abteilungs- und Klassenverhältnisses der 2 Worthälften die ersten 2 Zahlen auf Seite 5 fest.

Beispiel: „**linawzadem**" wird folgendermassen übersetzt:

Der Index verweist auf
Seite A 30 für die **1te** Worthälfte
» A 20 » » **2te** »

1te Worthälfte {**li** = 36, **naw** = 54} Abt. A, 4te Kl.
2te Worthälfte {**za** = 24, **dem** = 13} » A, 4te Kl.

laut Seite 5 ergeben sich aus:
1te Worthälfte A, 4te Klasse
und 2te » A, 4te »
als erste 2 Zahlen: 03.
linawzadem daher = 03 36 54 24 13

Beim **Übersetzen** einer **Zahl** in ein **Wort** stellt man auf Seite 5 mittels der ersten 2 Zahlen das Abteilungs- und Klassenverhältnis der 2 Worthälften fest; der Randvermerk erleichtert das Auffinden der Seiten, wo die
3, 4, 5, 6te Zahl und 1te Worthälfte
7, 8, 9, 10te » » **2te** »
zu suchen sind.

Beispiel: 21 23 46 98 83 ist in ein Wort zu übersetzen:

21 = {
1te Worthälfte B, 10te Kl.
(auf Seite B 20 zu suchen, da 3te Zahl eine 2.)
2te Worthälfte A, 10te Kl.
(auf Seite A 90 zu suchen, da 7te Zahl eine 9.)
}

23 = **ok**, 46 = **lus** } 1te Worthälfte
98 = **gu**, 83 = **tyz** } 2te Worthälfte

21 23 46 98 83 daher = **oklusgutyz.**

Hamburg, Juni 1907.

R. Vollers.

The numbering of the 60 pages A 00, A 10 etc. at the margin is very useful, since the number at the margin coincides with the 1st two horizontal upper figures.

Directions for the use of this system.

In translating a **word** into a **figure** group refer first to Index in order to find by the aid of the marginal numbering:
the prefix with 3rd, 4, 5 & 6th figures
» suffix » 7, 8, 9 » 10th »
(the horizontal letter groups are classed by their vowels)
and then look up page 5, where you will find the 1st and 2nd figures according to the section(s) (A, B, C) and class to which the prefix and suffix belong.

Example: "**linawzadem**" is to be translated as follows:

The Index will refer you to
page A 30 for the prefix
and » A 20 » » suffix

prefix {**li** = 36, **naw** = 54} Sect. A, 4th class
suffix {**za** = 24, **dem** = 13} » A, 4th »

as the prefix is in Sect. A, 4th class
and » suffix » » » A, 4th »
the 1st and 2nd figures as per page 5 must be: 03; **linawzadem** therefore = 03 36 54 24 13.

In **translating** a **figure** into a **word** refer first to table on page 5 to find out to which section(s) (A, B, C) and class the 2 half-words belong. The marginal numbering will lead you to the pages where you will find through the 3rd, 4th, 5th and 6th figures the **prefix**
and through the 7th, 8th, 9th and 10th figures the **suffix**

Example: 21 23 46 98 83 to be translated into a word:

21 = {
prefix B, 10th class
(the 3rd figure being a 2, the prefix must be on page B 20)
suffix A, 10th class
(the 7th figure being a 9, the suffix must be on page A 90)
}

23 = **ok**, 46 = **lus** } prefix
98 = **gu**, 83 = **tyz** } suffix

21 23 46 98 83 = "**oklusgutyz**:"

Hamburg, June 1907.

R. Vollers.

Système "Vollers" de 10 chiffres

10 000 Millions de mots de 10 lettres compilés avec la difference d'au moins 2 lettres et d'après les décisions du Còngrès télégraphique de Lòndres de l'année 1903. — Groupes de chiffres 0 000 000 000 à 9 999 999 999.

Le nombre total se compose de **12 classes** chacune de **900 Millions = 10 800 Millions** de mots entiers de 10 lettres; il a été réduit au nombre rond de **10 000** millions, comme on le verra à la page 5.

Chacune de ces 12 classes a 30 000 **demi**-mots de 5 lettres (10 000 A, 10 000 B, 10 000 C) qui peuvent former aussi bien la **première** que la **seconde** moitié des mots entiers et qui représentent ainsi 30 000 × 30 000 = 900 Millions de mots **complets** de 10 lettres.

Les 30 000 demi-mots de chaque classe ont entre eux la différence d'au moins **2** lettres et ne présentent aucune ressemblance concernant les symboles de "Morse" comme il en existerait par exemple entre **bn** —...|—. et **dr** —..|.—. **ainsi que nf** —.|..—.

La **première** et la **seconde moitié** de **chaque mot doivent appartenir à la même classe;** si non, la différence de **2** lettres se réduirait à 1 lettre; car il va sans dire que **tous** les **demi-mots** (360 000) existant dans ce livre n'ont entre eux que la difference d'au moins **1** lettre.

Les mots entiers étant harmonieux et facilement prononçables, se prêtent fort bien à l'usage de codes.

Les syllabes suivantes n'existent pas dans mes mots entiers:

1) **jj, xx, yy, iy, yll,**
2) **aaa, bbb, ccc,** à savoir: la triplication d'une lettre excepté **"s."**

Les groupes de 10 chiffres se composent comme suit: Les **8 derniers chiffres** résultent de la 1re et de la 2de moitié des mots et les **2 premiers** chiffres de la manière de composition de ces 2 moitiés concernant section(s) (A,B,C) et classe.

Le numérotage des 60 pages (A 00, A 10 etc.) est très utile vu

Sistema "Vollers" de 10 cifras

10 000 Millones de palabras de 10 letras construidas con la diferencia minima de 2 letras y en conformidad con las decisiones del Congreso telegráfico de Londres del año 1903. — Grupos de cifras 0 000 000 000 hasta 9 999 999 999.

El número está compuesto de **12 clases,** cada una de **900 Millones = 10 800 Millones** de palabras completas de 10 letras; esto está reducido al número redondo de **10 000** Millones como se verá en la página 5.

Cada una de estas 12 clases tiene 30 000 **medias** palabras de 5 letras (10 000 A, B y C) que se pueden **anteponer** ó **posponer** entre si, resultando 30 000 × 30 000 = 900 Millones de palabras de 10 letras.

Las **30 000** medias palabras de cada clase difieren en **2** letras á lo menos y no presentan ninguna semejanza respecto á los caractéres telegráficos de "Morse", como seria por ejemplo entre **bn** —...|—. y **dr** —..|.—. **y tambien nf** —.|..—.

La **primera** y **segunda mitad** de **cada palabra deben pertenecer á la misma clase,** si no, la diferencia de **2** letras se reduciria á **1** letra puesto que **todas** las **mitades** de palabra (360 000) de este libro difieren solamente en **1** letra á lo menos.

Las palabras completas son harmoniosas, se pronuncian facilmente y por consiguiente se prestan muy bien al uso de códigos.

Las combinaciones siguientes de letras han sido omitidas:

1) **jj, xx, yy, iy, yll,**
2) **aaa, bbb, ccc, ddd,** es decir: la triplicación de una letra exceptuando **"s."**

Los grupos de 10 cifras se componen como sigue: las **ultimas 8 cifras** resultan de la primera y segunda mitad de palabra, y las **2 primeras** del modo de composición de las 2 mitades respecto á la(s) seccion(es) (A, B, C) y clase.

El modo de numeración de las páginas es muy útil, puesto que el número del margen indica las

que le nombre en marge indique les 2 premiers chiffres horizontaux de chaque page.

Mode d'emploi.

En **traduisant** un **mot** en **chiffres** il faut chercher d'abord à l'aide de l'index et puis par le numérotage à la marge des pages en question:

la 1re moitié du mot avec les 3me, 4, 5 et 6mes chiffres et
» 2de » du mot avec les 7, 8, 9 et 10nes chiffres

(*les syllabes horizontales sont classées par leurs voyelles*)

Il reste à trouver les 2 premiers chiffres à la page 5 d'après les(la) Section(s) (A, B, C) et la classe auxquelles les 2 demi-mots appartiennent.

Exemple: Traduisons le **mot** **"linawzadem"**: d'après l'index le
1er demi-mot est à la page A 30 et
2d » » » » » » A 20

1er demi-mot { **li** = 36, **naw** = 54 } Section A, 4e classe
2d demi-mot { **za** = 24, **dem** = 13 } Section A, 4e classe

le 1er demi-mot se trouvant à la Section A, 4e cl.
» 2d » » » A, 4e »

les 2 premiers chiffres d'après la page 5 doivent être: **03.**
donc **"linawzadem"** = 03 36 54 24 13.

En **traduisant** un **groupe de chiffres** en un **mot** on traduira d'abord à la page 5 les 2 premiers chiffres pour constater à quelle(s) section(s)(A,B,C) et à quelle classe les moitiés du mot appartiennent. Le numérotage en marge vous indiquera les pages où vous trouverez

les 3, 4, 5 et 6mes chiffres avec le premier demi-mot et
» 7, 8, 9 et 10mes chiffres avec le dernier demi-mot.

Exemple: Traduisons 21 23 46 98 83 en un **mot:**

21 = { 1er demi-mot B, 10me classe (*le 3me chiffre étant 2, cherchez ce demi-mot page B 20*); 2nd demi-mot A, 10me classe (*le 7me chiffre étant 9, cherchez ce demi-mot page A 90*) }

23 = **ok**, 46 = **lus** } 1er demi-mot
98 = **gu**, 83 = **tyz** } 2d » »

donc: **21 23 46 98 83** = **oklusgutyz.**

Hambourg, Juin 1907.

R. Vollers.

2 primeras cifras horizontales de cada pagina.

Modo de empleo.

Traduciendo una **palabra** en un **grupo de cifras** se debe buscar primero mediante el "Index" y despues mediante la numeración al margen de las páginas en cuestión:

la 1ª media-palabra con las 3ª, 4ª, 5ª, 6ª cifras y
» 2ª » » con las 7ª, 8ª, 9ª, 10ª cifras

(*las silabas horizontales están agrupadas por sus vocales*);

ahora resta hallar las 2 primeras cifras en la página 5 según la(s) seccion(es) (A, B, C) y la clase á las cuales las 2 mitades pertenecen.

Ejemplo: Traducimos la palabra **"linawzadem"**: Según el Index:
la 1ª media-palabra debe hallarse en la pág. A 30
» 2ª » » » » » A 20

1ª media-palabra { **li** = 36, **naw** = 54 } Seccion A, 4ª clase
2ª media-palabra { **za** = 24, **dem** = 13 } Seccion A, 4ª clase

la 1ª media-palabra estando en la Sección A, 4ª cl.
» 2ª » » » A, 4ª »

las 2 primeras cifras según pág. 5 deben estar: **03.** Entonces:
"linawzadem" = 03 36 54 24 13.

Traduciendo un **grupo de cifras** en **una palabra** se traducirá primeramente en la pág. 5 las 2 primeras cifras para encontrar á qué seccion(es) (A, B, C) y á qué clase las 2 mitades pertenecen. Los números en el margen indicarán las páginas donde se hallan:

las 3ª, 4ª, 5ª y 6ª cifras con la 1ª media-palabra y
» 7ª, 8ª, 9ª y 10ª cifras con la 2ª media-palabra.

Ejemplo: Traducimos 21 23 46 98 83 en una palabra:

21 = { 1ª media-palabra B, 10ª cl. (*siendo la 3a cifra 2, busquese esta media-palabra en la página B 20*); 2ª media-palabra A, 10ª cl. (*siendo la 7a cifra 9 busquese esta media-palabra en la página A 90*) }

23 = **ok**, 46 = **lus** } 1ª media-palabra
98 = **gu**, 83 = **tyz** } 2ª » »

entonces 21 23 46 98 83 = **oklusgutyz**

Hamburg, Junio 1907.

R. Vollers.

The 1st and 2nd figures

of the complete group of 10 figures result from the manner of composition of the 2 half-words (prefix and suffix) as regards section(s) (A B C) and class as per following table:

		and **last** half-word (suffix)		
		A same class	**B** same class	**C** same class
		First 2 figures	First 2 figures	First 2 figures
First half-word (prefix)	**A** 1st class	00	36	72
	» 2nd »	01	37	73
	» 3rd »	02	38	74
	» 4th »	03	39	75
	» 5 » »	04	40	76
	» 6 » »	05	41	77
	» 7 » »	06	42	78
	« 8 » »	07	43	79
	» 9 » »	08	44	80
	» 10 » »	09	45	81
	» 11 » »	10	46	82
	» 12 » »	11	47	83
	B 1st class	12	48	84
	» 2nd »	13	49	85
	» 3rd »	14	50	86
	» 4th »	15	51	87
	» 5 » »	16	52	88
	» 6 » »	17	53	89
	» 7 » »	18	54	90
	» 8 » »	19	55	91
	» 9 » »	20	56	92
	» 10 » »	21	57	93
	» 11 » »	22	58	94
	» 12 » »	23	59	95
	C 1st class	24	60	96
	» 2nd »	25	61	97
	» 3rd »	26	62	98
	» 4th »	27	63	99
	» 5 » »	28	64	
	» 6 » »	29	65	
	» 7 » »	30	66	
	» 8 » »	31	67	
	» 9 » »	32	68	
	» 10 » »	33	69	
	» 11 » »	34	70	
	» 12 » »	35	71	

INDEX *to be used only for the **translation** of telegrams*

for half-words beginning with a **vowel**

if the 3rd letter of the half-word is a vowel, refer to ***Section C.***

a	e	i	o	u	
A or C 00	A or C 70	A or C 20	B or C 10	B or C 60	**b**
» 30	B or C 20	B or C 40	» 90	» 70	**c**
» 10	» 00	A or C 50	» 70	A or C 20	**d**
» 20	» 40	B or C 30	A or C 50	» 60	**f**
B or C 80	» 30	» 20	» 10	» 70	**g**
A or C 60	A or C 90		B or C 00		**j**
» 40	B or C 60	» 90	» 20	B or C 80	**k**
» 70	A or C 80	A or C 40	» 30	» 90	**l**
B or C 10	» 10	» 70	» 60	A or C 00	**m**
A or C 90	B or C 50	» 80	A or C 00	B or C 30	**n**
» 80	» 70	» 00	B or C 80	A or C 10	**p**
B or C 40	» 90	» 90	A or C 80	» 40	**r**
» 30	A or C 40	B or C 60	B or C 40	B or C 00	**s**
» 50	» 30	A or C 60	A or C 70	A or C 80	**t**
» 70	» 20	B or C 10	» 40	» 90	**v**
» 60		» 50	» 30	B or C 20	**w**
» 20	» 50	» 80	» 60	A or C 30	**x**
» 00	B er C 10	A or C 10	B or C 50	» 50	**z**

for half-words beginning with a **consonant**

if the 3rd letter of the half-word is a vowel, refer to ***Section C.***

b	c	d	f	g	h	j	k	l	m	n	p	r	s	t	v	w	x	z	
B C 50	B C 20	B C 60	B C 70	B C 40	B C 90	A C 00	B C 80	B C 10	A C 30	A C 40	A C 50	A C 60	A C 70	A C 80	B C 00	A C 90	A C 10	A C 20	**a**
B C 30	B C 00	B C 40	A C 70	A C 10	B C 70	B C 90	A C 00	A C 40	A C 20	A C 50	B C 50	A C 80	A C 60	B C 80	A C 30	B C 60	B C 10	A C 90	**e**
A C 70	B C 90	B C 20	A C 90	B C 60	A. C 20	A C 40	A C 50	A C 30	B C 70	A C 10	B C 40	A C 00	B C 80	B C 10	B C 50	B C 30	B C 00	A C 60	**i**
A C 10	B C 80	B C 00	B C 10	B C 20	A C 30	A C 50	B C 50	A C 60	B C 40	B C 60	A C 90	B C 70	A C 40	A C 20	A C 00	B C 90	A C 80	B C 30	**o**
B C 80	A C 20	B C 90	A C 50	A C 90	A C 00	B C 70	A C 70	B C 20	B C 50	B C 40	A C 40	A C 30	B C 30	B C 00	A C 80	B C 10	A C 60	B C 60	**u**
A C 60	B C 40	A C 80	B C 00	A C 50		A C 20	B C 10	B C 50	A C 90	B C 20	B C 30	A C 70	A C 10	B C 70	B C 80	A C 00	A C 30	B C 90	**y**

Examples:

ce : page B 00 or page C 00

lg : page B 20 or page C 20

by : page A 60 or page C 60

Further examples. Weitere Beispiele.
D'autres exemples. Otros ejemplos.

	at	aly	fo	def
68	50	56	17	13

C & B
9 cl.

atalyfodef = 6850561713

	ry	cec	mo	afu
88	79	07	47	51

A & C
12 cl.

rycecmoafu = 8379074751

	ok	mer	vi	ubo
94	23	49	56	90

B & C
11 cl.

okmerviubo = 9423495690

	en	epy	le	idy
99	51	66	46	21

C & C
4 cl.

enepyleidy = 9951664621

	fe	sel	ot	son
05	76	73	73	75

A & A
6 cl.

feselotson = 0576737375

	am	pia	wu	das
48	10	62	18	12

B & B
1 cl.

ampiawudas = 4810621812

A

First 2 of the 5 letters	ab	ip	on	um	ja	ke	ri	vo	hu	wy
	00	01	02	03	04	05	06	07	08	09

last 3 of the 5 letters:

	1st	2nd	3rd	4th	5th	6th		7th	8th	9th	10th	11th	12th class
A 00	bac	bad	baf	bag	bah	bai	**00**	bak	bal	bam	ban	bao	bap
	bed	bef	beg	beh	bea	bek	**01**	bel	bem	ben	beo	bep	ber
	bif	big	bih	bia	bik	bil	**02**	bim	bin	bio	bip	bir	bis
	bog	boh	boa	bok	bol	bom	**03**	bon	boy	bop	bor	bos	bot
	buh	bua	buk	bul	bum	bun	**04**	buo	bup	bur	bus	but	buv
	bya	byk	byl	bym	byn	byo	**05**	byp	byr	bys	byt	byv	byw
	cad	caf	cag	cah	cai	cak	**06**	cal	cam	can	cao	cap	car
	cef	ceg	ceh	cea	cek	cel	**07**	cem	cen	ceo	cep	cer	ces
	cig	cih	cia	cik	cil	cim	**08**	cin	cio	cip	cir	cis	cit
	coh	coa	cok	col	com	con	**09**	coy	cop	cor	cos	cot	cov
	cua	cuk	cul	cum	cun	cuo	**10**	cup	cur	cus	cut	cuv	cuw
	cyk	cyl	cym	cyn	cyo	cyp	**11**	cyr	cys	cyt	cyv	cyw	cye
	daf	dag	dah	dai	dak	dal	**12**	dam	dan	dao	dap	dar	das
	deg	deh	dea	dek	del	dem	**13**	den	deo	dep	der	des	det
	dih	dia	dik	dil	dim	din	**14**	dio	dip	dir	dis	dit	div
	doa	dok	dol	dom	don	doy	**15**	dop	dor	dos	dot	dov	dow
	duk	dul	dum	dun	duo	dup	**16**	dur	dus	dut	duv	duw	due
	dyl	dym	dyn	dyo	dyp	dyr	**17**	dys	dyt	dyv	dyw	dye	dyz
	fag	fah	fai	fak	fal	fam	**18**	fan	fao	fap	far	fas	fat
	feh	fea	fek	fel	fem	fen	**19**	feo	fep	fer	fes	fet	fev
	fia	fik	fil	fim	fin	fio	**20**	fip	fir	fis	fit	fiv	fiw
	fok	fol	fom	fon	foy	fop	**21**	for	fos	fot	fov	fow	foe
	ful	fum	fun	fuo	fup	fur	**22**	fus	fut	fuv	fuw	fue	fuz
	fym	fyn	fyo	fyp	fyr	fys	**23**	fyt	fyv	fyw	fye	fyz	fyb
	gah	gai	gak	gal	gam	gan	**24**	gao	gap	gar	gas	gat	gav
	gea	gek	gel	gem	gen	geo	**25**	gep	ger	ges	get	gev	gey
	gik	gil	gim	gin	gio	gip	**26**	gir	gis	git	giv	giw	gie
	gol	gom	gon	goy	gop	gor	**27**	gos	got	gov	gow	goe	goz
	gum	gun	guo	gup	gur	gus	**28**	gut	guv	guw	gue	guz	gub
	gyn	gyo	gyp	gyr	gys	gyt	**29**	gyv	gyw	gye	gyz	gyb	gyc
	hai	hak	hal	ham	han	hao	**30**	hap	har	has	hat	hav	haw
	hek	hel	hem	hen	heo	hep	**31**	her	hes	het	hev	hey	heu
	hil	him	hin	hio	hip	hir	**32**	his	hit	hiv	hiw	hie	hiz
	hom	hon	hoy	hop	hor	hos	**33**	hot	hov	how	hoe	hoz	hob
	hun	huo	hup	hur	hus	hut	**34**	huv	huw	hue	huz	hub	huc
	hyo	hyp	hyr	hys	hyt	hyv	**35**	hyw	hye	hyz	hyb	hyc	hyd
	kal	kam	kan	kao	kap	kar	**36**	kas	kat	kav	kaw	kae	kaz
	kem	ken	keo	kep	ker	kes	**37**	ket	kev	key	keu	kez	keb
	kin	kio	kip	kir	kis	kit	**38**	kiv	kiw	kie	kiz	kib	kic
	koy	kop	kor	kos	kot	kov	**39**	kow	koe	koz	kob	koc	kod
	kup	kur	kus	kut	kuv	kuw	**40**	kue	kuz	kub	kuc	kud	kuf
	kyr	kys	kyt	kyv	kyw	kye	**41**	kyz	kyb	kyc	kyd	kyf	kyg
	lam	lan	lao	lap	lar	las	**42**	lat	lav	law	lae	laz	lab
	len	leo	lep	ler	les	let	**43**	lev	ley	leu	lez	leb	lec
	lio	lip	lir	lis	lit	liv	**44**	liw	lie	liz	lib	lic	lid
	lop	lor	los	lot	lov	low	**45**	loe	loz	lob	loc	lod	lof
	lur	lus	lut	luv	luw	lue	**46**	luz	lub	luc	lud	luf	lug
	lys	lyt	lyv	lyw	lye	lyz	**47**	lyb	lyc	lyd	lyf	lyg	lyh
	man	mao	map	mar	mas	mat	**48**	mav	maw	mae	maz	mab	mac
	meo	mep	mer	mes	met	mev	**49**	mey	meu	mez	meb	mec	med

A

First 2 of the 5 letters	ab	ip	on	um	ja	ke	ri	vo	hu	wy
	00	01	02	03	04	05	06	07	08	09

last 3 of the 5 letters:

A 00

1st	2nd	3rd	4th	5th	6th		7th	8th	9th	10th	11th	12th class
mip	mir	mis	mit	miv	miw	**50**	mie	miz	mib	mic	mid	mif
mor	mos	mot	mov	mow	moe	**51**	moz	mob	moc	mod	mof	mog
mus	mut	muv	muw	mue	muz	**52**	mub	muc	mud	muf	mug	muh
myt	myv	myw	mye	myz	myb	**53**	myc	myd	myf	myg	myh	mya
nao	nap	nar	nas	nat	nav	**54**	naw	nae	naz	nab	nac	nad
nep	ner	nes	net	nev	ney	**55**	neu	nez	neb	nec	ned	nef
nir	nis	nit	niv	niw	nie	**56**	niz	nib	nic	nid	nif	nig
nos	not	nov	now	noe	noz	**57**	nob	noc	nod	nof	nog	noh
nut	nuv	nuw	nue	nuz	nub	**58**	nuc	nud	nuf	nug	nuh	nua
nyv	nyw	nye	nyz	nyb	nyc	**59**	nyd	nyf	nyg	nyh	nya	nyk
par	pas	pat	pav	paw	pae	**60**	paz	pab	pac	pad	paf	pag
pes	pet	pev	pey	peu	pez	**61**	peb	pec	ped	pef	peg	peh
pit	piv	piw	pie	piz	pib	**62**	pic	pid	pif	pig	pih	pia
pov	pow	poe	poz	pob	poc	**63**	pod	pof	pog	poh	poa	pok
puw	pue	puz	pub	puc	pud	**64**	puf	pug	puh	pua	puk	pul
pye	pyz	pyb	pyc	pyd	pyf	**65**	pyg	pyh	pya	pyk	pyl	pym
ras	rat	rav	raw	rae	raz	**66**	rab	rac	rad	raf	rag	rah
ret	rev	rey	reu	rez	reb	**67**	rec	red	ref	reg	reh	rea
riv	riw	rie	riz	rib	ric	**68**	rid	rif	rig	rih	ria	rik
row	roe	roz	rob	roc	rod	**69**	rof	rog	roh	roa	rok	rol
rue	ruz	rub	ruc	rud	ruf	**70**	rug	ruh	rua	ruk	rul	rum
ryz	ryb	ryc	ryd	ryf	ryg	**71**	ryh	rya	ryk	ryl	rym	ryn
sat	sav	saw	sae	saz	sab	**72**	sac	sad	saf	sag	sah	sai
sev	sey	seu	sez	seb	sec	**73**	sed	sef	seg	seh	sea	sek
siw	sie	siz	sib	sic	sid	**74**	sif	sig	sih	sia	sik	sil
soe	soz	sob	soc	sod	sof	**75**	sog	soh	soa	sok	sol	som
suz	sub	suc	sud	suf	sug	**76**	suh	sua	suk	sul	sum	sun
syb	syc	syd	syf	syg	syh	**77**	sya	syk	syl	sym	syn	syo
tav	taw	tae	taz	tab	tac	**78**	tad	taf	tag	tah	tai	tak
tey	teu	tez	teb	tec	ted	**79**	tef	teg	teh	tea	tek	tel
tie	tiz	tib	tic	tid	tif	**80**	tig	tih	tia	tik	til	tim
toz	tob	toc	tod	tof	tog	**81**	toh	toa	tok	tol	tom	ton
tub	tuc	tud	tuf	tug	tuh	**82**	tua	tuk	tul	tum	tun	tuo
tyc	tyd	tyf	tyg	tyh	tya	**83**	tyk	tyl	tym	tyn	tyo	typ
vaw	vae	vaz	vab	vac	vad	**84**	vaf	vag	vah	vai	vak	val
veu	vez	veb	vec	ved	vef	**85**	veg	veh	vea	vek	vel	vem
viz	vib	vic	vid	vif	vig	**86**	vih	via	vik	vil	vim	vin
vob	voc	vod	vof	vog	voh	**87**	voa	vok	vol	vom	von	voy
vuc	vud	vuf	vug	vuh	vua	**88**	vuk	vul	vum	vun	vuo	vup
vyd	vyf	vyg	vyh	vya	vyk	**89**	vyl	vym	vyn	vyo	vyp	vyr
wae	waz	wab	wac	wad	waf	**90**	wag	wah	wai	wak	wal	wam
wez	web	wec	wed	wef	weg	**91**	weh	wea	wek	wel	wem	wen
wib	wic	wid	wif	wig	wih	**92**	wia	wik	wil	wim	win	wio
woc	wod	wof	wog	woh	woa	**93**	wok	wol	wom	won	woy	wop
zab	zac	zad	zaf	zag	zah	**94**	zai	zak	zal	zam	zan	zao
zec	zed	zef	zeg	zeh	zea	**95**	zek	zel	zem	zen	zeo	zep
zid	zif	zig	zih	zia	zik	**96**	zil	zim	zin	zio	zip	zir
zof	zog	zoh	zoa	zok	zol	**97**	zom	zon	zoy	zop	zor	zos
zug	zuh	zua	zuk	zul	zum	**98**	zun	zuo	zup	zur	zus	zut
zyh	zya	zyk	zyl	zym	zyn	**99**	zyo	zyp	zyr	zys	zyt	zyv

A

First 2 of the 5 letters:	ad	em	iz	og	up	xa	ge	ni	bo	sy
last 3 of the 5 letters:	10	11	12	13	14	15	16	17	18	19

	1st	2nd	3rd	4th	5th	6th		7th	8th	9th	10th	11th	12th class
A	bad	baf	bag	bah	bai	bak	00	bal	bam	ban	bao	bap	bar
	bef	beg	beh	bea	bek	bel	01	bem	ben	beo	bep	ber	bes
10	big	bih	bia	bik	bil	bim	02	bin	bio	bip	bir	bis	bit
	boh	boa	bok	bol	bom	bon	03	boy	bop	bor	bos	bot	bov
	bua	buk	bul	bum	bun	buo	04	bup	bur	bus	but	buv	buw
	byk	byl	bym	byn	byo	byp	05	byr	bys	byt	byv	byw	bye
	caf	cag	cah	cai	cak	cal	06	cam	can	cao	cap	car	cas
	ceg	ceh	cea	cek	cel	cem	07	cen	ceo	cep	cer	ces	cet
	cih	cia	cik	cil	cim	cin	08	cio	cip	cir	cis	cit	civ
	coa	cok	col	com	con	coy	09	cop	cor	cos	cot	cov	cow
	cuk	cul	cum	cun	cuo	cup	10	cur	cus	cut	cuv	cuw	cue
	cyl	cym	cyn	cyo	cyp	cyr	11	cys	cyt	cyv	cyw	cye	cyz
	dag	dah	dai	dak	dal	dam	12	dan	dao	dap	dar	das	dat
	deh	dea	dek	del	dem	den	13	deo	dep	der	des	det	dev
	dia	dik	dil	dim	din	dio	14	dip	dir	dis	dit	div	diw
	dok	dol	dom	don	doy	dop	15	dor	dos	dot	dov	dow	doe
	dul	dum	dun	duo	dup	dur	16	dus	dut	duv	duw	due	duz
	dym	dyn	dyo	dyp	dyr	dys	17	dyt	dyv	dyw	dye	dyz	dyb
	fah	fai	fak	fal	fam	fan	18	fao	fap	far	fas	fat	fav
	fea	fek	fel	fem	fen	feo	19	fep	fer	fes	fet	fev	fey
	fik	fil	fim	fin	fio	fip	20	fir	fis	fit	fiv	fiw	fie
	fol	fom	fon	foy	fop	for	21	fos	fot	fov	fow	foe	foz
	fum	fun	fuo	fup	fur	fus	22	fut	fuv	fuw	fue	fuz	fub
	fyn	fyo	fyp	fyr	fys	fyt	23	fyv	fyw	fye	fyz	fyb	fyc
	gai	gak	gal	gam	gan	gao	24	gap	gar	gas	gat	gav	gaw
	gek	gel	gem	gen	geo	gep	25	ger	ges	get	gev	gey	geu
	gil	gim	gin	gio	gip	gir	26	gis	git	giv	giw	gie	giz
	gom	gon	goy	gop	gor	gos	27	got	gov	gow	goe	goz	gob
	gun	guo	gup	gur	gus	gut	28	guv	guw	gue	guz	gub	guc
	gyo	gyp	gyr	gys	gyt	gyv	29	gyw	gye	gyz	gyb	gyc	gyd
	hak	hal	ham	han	hao	hap	30	har	has	hat	hav	haw	hae
	hel	hem	hen	heo	hep	her	31	hes	het	hev	hey	heu	hez
	him	hin	hio	hip	hir	his	32	hit	hiv	hiw	hie	hiz	hib
	hon	hoy	hop	hor	hos	hot	33	hov	how	hoe	hoz	hob	hoc
	huo	hup	hur	hus	hut	huv	34	huw	hue	huz	hub	huc	hud
	hyp	hyr	hys	hyt	hyv	hyw	35	hye	hyz	hyb	hyc	hyd	hyf
	kam	kan	kao	kap	kar	kas	36	kat	kav	kaw	kae	kaz	kab
	ken	keo	kep	ker	kes	ket	37	kev	key	keu	kez	keb	kec
	kio	kip	kir	kis	kit	kiv	38	kiw	kie	kiz	kib	kic	kid
	kop	kor	kos	kot	kov	kow	39	koe	koz	kob	koc	kod	kof
	kur	kus	kut	kuv	kuw	kue	40	kuz	kub	kuc	kud	kuf	kug
	kys	kyt	kyv	kyw	kye	kyz	41	kyb	kyc	kyd	kyf	kyg	kyh
	lan	lao	lap	lar	las	lat	42	lav	law	lae	laz	lab	lac
	leo	lep	ler	les	let	lev	43	ley	leu	lez	leb	lec	led
	lip	lir	lis	lit	liv	liw	44	lie	liz	lib	lic	lid	lif
	lor	los	lot	lov	low	loe	45	loz	lob	loc	lod	lof	log
	lus	lut	luv	luw	lue	luz	46	lub	luc	lud	luf	lug	luh
	lyt	lyv	lyw	lye	lyz	lyb	47	lyc	lyd	lyf	lyg	lyh	lya
	mao	map	mar	mas	mat	mav	48	maw	mae	maz	mab	mac	mad
	mep	mer	mes	met	mev	mey	49	meu	mez	meb	mec	med	mef

A

First 2 of the 5 letters:	ad	em	iz	og	up	xa	ge	ni	bo	sy
	10	11	12	13	14	15	16	17	18	19

last 3 of the 5 letters:

1st	2nd	3rd	4th	5th	6th		7th	8th	9th	10th	11th	12th class
mir	mis	mit	miv	miw	mie	**50**	miz	mib	mic	mid	mif	mig
mos	mot	mov	mow	moe	moz	**51**	mob	moc	mod	mof	mog	moh
mut	muv	muw	mue	muz	mub	**52**	muc	mud	muf	mug	muh	mua
myv	myw	mye	myz	myb	myc	**53**	myd	myf	myg	myh	mya	myk
nap	nar	nas	nat	nav	naw	**54**	nae	naz	nab	nac	nad	naf
ner	nes	net	nev	ney	neu	**55**	nez	neb	nec	ned	nef	neg
nis	nit	niv	niw	nie	niz	**56**	nib	nic	nid	nif	nig	nih
not	nov	now	noe	noz	nob	**57**	noc	nod	nof	nog	noh	noa
nuv	nuw	nue	nuz	nub	nuc	**58**	nud	nuf	nug	nuh	nua	nuk
nyw	nye	nyz	nyb	nyc	nyd	**59**	nyf	nyg	nyh	nya	nyk	nyl
pas	pat	pav	paw	pae	paz	**60**	pab	pac	pad	paf	pag	pah
pet	pev	pey	peu	pez	peb	**61**	pec	ped	pef	peg	peh	pea
piv	piw	pie	piz	pib	pic	**62**	pid	pif	pig	pih	pia	pik
pow	poe	poz	pob	poc	pod	**63**	pof	pog	poh	poa	pok	pol
pue	puz	pub	puc	pud	puf	**64**	pug	puh	pua	puk	pul	pum
pyz	pyb	pyc	pyd	pyf	pyg	**65**	pyh	pya	pyk	pyl	pym	pyn
rat	rav	raw	rae	raz	rab	**66**	rac	rad	raf	rag	rah	rai
rev	rey	reu	rez	reb	rec	**67**	red	ref	reg	reh	rea	rek
riw	rie	riz	rib	ric	rid	**68**	rif	rig	rih	ria	rik	ril
roe	roz	rob	roc	rod	rof	**69**	rog	roh	roa	rok	rol	rom
ruz	rub	ruc	rud	ruf	rug	**70**	ruh	rua	ruk	rul	rum	run
ryb	ryc	ryd	ryf	ryg	ryh	**71**	rya	ryk	ryl	rym	ryn	ryo
sav	saw	sae	saz	sab	sac	**72**	sad	saf	sag	sah	sai	sak
sey	seu	sez	seb	sec	sed	**73**	sef	seg	seh	sea	sek	sel
sie	siz	sib	sic	sid	sif	**74**	sig	sih	sia	sik	sil	sim
soz	sob	soc	sod	sof	sog	**75**	soh	soa	sok	sol	som	son
sub	suc	sud	suf	sug	suh	**76**	sua	suk	sul	sum	sun	suo
syc	syd	syf	syg	syh	sya	**77**	syk	syl	sym	syn	syo	syp
taw	tae	taz	tab	tac	tad	**78**	taf	tag	tah	tai	tak	tal
teu	tez	teb	tec	ted	tef	**79**	teg	teh	tea	tek	tel	tem
tiz	tib	tic	tid	tif	tig	**80**	tih	tia	tik	til	tim	tin
tob	toc	tod	tof	tog	toh	**81**	toa	tok	tol	tom	ton	toy
tuc	tud	tuf	tug	tuh	tua	**82**	tuk	tul	tum	tun	tuo	tup
tyd	tyf	tyg	tyh	tya	tyk	**83**	tyl	tym	tyn	tyo	typ	tyr
vae	vaz	vab	vac	vad	vaf	**84**	vag	vah	vai	vak	val	vam
vez	veb	vec	ved	vef	veg	**85**	veh	vea	vek	vel	vem	ven
vib	vic	vid	vif	vig	vih	**86**	via	vik	vil	vim	vin	vio
voc	vod	vof	vog	voh	voa	**87**	vok	vol	vom	von	voy	vop
vud	vuf	vug	vuh	vua	vuk	**88**	vul	vum	vun	vuo	vup	vur
vyf	vyg	vyh	vya	vyk	vyl	**89**	vym	vyn	vyo	vyp	vyr	vys
waz	wab	wac	wad	waf	wag	**90**	wah	wai	wak	wal	wam	wan
web	wec	wed	wef	weg	weh	**91**	wea	wek	wel	wem	wen	weo
wic	wid	wif	wig	wih	wia	**92**	wik	wil	wim	win	wio	wip
wod	wof	wog	woh	woa	wok	**93**	wol	wom	won	woy	wop	wor
zac	zad	zaf	zag	zah	zai	**94**	zak	zal	zam	zan	zao	zap
zed	zef	zeg	zeh	zea	zek	**95**	zel	zem	zen	zeo	zep	zer
zif	zig	zih	zia	zik	zil	**96**	zim	zin	zio	zip	zir	zis
zog	zoh	zoa	zok	zol	zom	**97**	zon	zoy	zop	zor	zos	zot
zuh	zua	zuk	zul	zum	zun	**98**	zuo	zup	zur	zus	zut	zuv
zya	zyk	zyl	zym	zyn	zyo	**99**	zyp	zyr	zys	zyt	zyv	zyw

A 10

A

First 2 of the 5 letters: af 20 · ev 21 · ib 22 · ud 23 · za 24 · me 25 · hi 26 · to 27 · cu 28 · jy 29

last 3 of the 5 letters:

A 20

1st	2nd	3rd	4th	5th	6th		7th	8th	9th	10th	11th	12th class
baf	bag	bah	bai	bak	bal	00	bam	ban	bao	bap	bar	bas
beg	beh	bea	bek	bel	bem	01	ben	beo	bep	ber	bes	bet
bih	bia	bik	bil	bim	bin	02	bio	bip	bir	bis	bit	biv
boa	bok	bol	bom	bon	boy	03	bop	bor	bos	bot	bov	bow
buk	bul	bum	bun	buo	bup	04	bur	bus	but	buv	buw	bue
byl	bym	byn	byo	byp	byr	05	bys	byt	byv	byw	bye	byz
cag	cah	cai	cak	cal	cam	06	can	cao	cap	car	cas	cat
ceh	cea	cek	cel	cem	cen	07	ceo	cep	cer	ces	cet	cev
cia	cik	cil	cim	cin	cio	08	cip	cir	cis	cit	civ	ciw
cok	col	com	con	coy	cop	09	cor	cos	cot	cov	cow	coe
cul	cum	cun	cuo	cup	cur	10	cus	cut	cuv	cuw	cue	cuz
cym	cyn	cyo	cyp	cyr	cys	11	cyt	cyv	cyw	cye	cyz	cyb
dah	dai	dak	dal	dam	dan	12	dao	dap	dar	das	dat	dav
dea	dek	del	dem	den	deo	13	dep	der	des	det	dev	dey
dik	dil	dim	din	dio	dip	14	dir	dis	dit	div	diw	die
dol	dom	don	doy	dop	dor	15	dos	dot	dov	dow	doe	doz
dum	dun	duo	dup	dur	dus	16	dut	duv	duw	due	duz	dub
dyn	dyo	dyp	dyr	dys	dyt	17	dyv	dyw	dye	dyz	dyb	dyc
fai	fak	fal	fam	fan	fao	18	fap	far	fas	fat	fav	faw
fek	fel	fem	fen	feo	fep	19	fer	fes	fet	fev	fey	feu
fil	fim	fin	fio	fip	fir	20	fis	fit	fiv	fiw	fie	fiz
fom	fon	foy	fop	for	fos	21	fot	fov	fow	foe	foz	fob
fun	fuo	fup	fur	fus	fut	22	fuv	fuw	fue	fuz	fub	fuc
fyo	fyp	fyr	fys	fyt	fyv	23	fyw	fye	fyz	fyb	fyc	fyd
gak	gal	gam	gan	gao	gap	24	gar	gas	gat	gav	gaw	gae
gel	gem	gen	geo	gep	ger	25	ges	get	gev	gey	geu	gez
gim	gin	gio	gip	gir	gis	26	git	giv	giw	gie	giz	gib
gon	goy	gop	gor	gos	got	27	gov	gow	goe	goz	gob	goc
guo	gup	gur	gus	gut	guv	28	guw	gue	guz	gub	guc	gud
gyp	gyr	gys	gyt	gyv	gyw	29	gye	gyz	gyb	gyc	gyd	gyf
hal	ham	han	hao	hap	har	30	has	hat	hav	haw	hae	haz
hem	hen	heo	hep	her	hes	31	het	hev	hey	heu	hez	heb
hin	hio	hip	hir	his	hit	32	hiv	hiw	hie	hiz	hib	hic
hoy	hop	hor	hos	hot	hov	33	how	hoe	hoz	hob	hoc	hod
hup	hur	hus	hut	huv	huw	34	hue	huz	hub	huc	hud	huf
hyr	hys	hyt	hyv	hyw	hye	35	hyz	hyb	hyc	hyd	hyf	hyg
kan	kao	kap	kar	kas	kat	36	kav	kaw	kae	kaz	kab	kac
keo	kep	ker	kes	ket	kev	37	key	keu	kez	keb	kec	ked
kip	kir	kis	kit	kiv	kiw	38	kie	kiz	kib	kic	kid	kif
kor	kos	kot	kov	kow	koe	39	koz	kob	koc	kod	kof	kog
kus	kut	kuv	kuw	kue	kuz	40	kub	kuc	kud	kuf	kug	kuh
kyt	kyv	kyw	kye	kyz	kyb	41	kyc	kyd	kyf	kyg	kyh	kya
lao	lap	lar	las	lat	lav	42	law	lae	laz	lab	lac	lad
lep	ler	les	let	lev	ley	43	leu	lez	leb	lec	led	lef
lir	lis	lit	liv	liw	lie	44	liz	lib	lic	lid	lif	lig
los	lot	lov	low	loe	loz	45	lob	loc	lod	lof	log	loh
lut	luv	luw	lue	luz	lub	46	luc	lud	luf	lug	luh	lua
lyv	lyw	lye	lyz	lyb	lyc	47	lyd	lyf	lyg	lyh	lya	lyk
map	mar	mas	mat	mav	maw	48	mae	maz	mab	mac	mad	maf
mer	mes	met	mev	mey	meu	49	mez	meb	mec	med	mef	meg

A

First 2 of the 5 letters: { **af** 20 | **ev** 21 | **ib** 22 | **ud** 23 | **za** 24 | **me** 25 | **hi** 26 | **to** 27 | **cu** 28 | **jy** 29

last 3 of the 5 letters:

1st	2nd	3rd	4th	5th	6th		7th	8th	9th	10th	11th	12th class	
mis	mit	miv	miw	mie	miz	50	mib	mic	mid	mif	mig	mih	A
mot	mov	mow	moe	moz	mob	51	moc	mod	mof	mog	moh	moa	
muv	muw	mue	muz	mub	muc	52	mud	muf	mug	muh	mua	muk	20
myw	mye	myz	myb	myc	myd	53	myf	myg	myh	mya	myk	myl	
nar	nas	nat	nav	naw	nae	54	naz	nab	nac	nad	naf	nag	
nes	net	nev	ney	neu	nez	55	neb	nec	ned	nef	neg	neh	
nit	niv	niw	nie	niz	nib	56	nic	nid	nif	nig	nih	nia	
nov	now	noe	noz	nob	noc	57	nod	nof	nog	noh	noa	nok	
nuw	nue	nuz	nub	nuc	nud	58	nuf	nug	nuh	nua	nuk	nul	
nye	nyz	nyb	nyc	nyd	nyf	59	nyg	nyh	nya	nyk	nyl	nym	
pat	pav	paw	pae	paz	pab	60	pac	pad	paf	pag	pah	pai	
pev	pey	peu	pez	peb	pec	61	ped	pef	peg	peh	pea	pek	
piw	pie	piz	pib	pic	pid	62	pif	pig	pih	pia	pik	pil	
poe	poz	pob	poc	pod	pof	63	pog	poh	poa	pok	pol	pom	
puz	pub	puc	pud	puf	pug	64	puh	pua	puk	pul	pum	pun	
pyb	pyc	pyd	pyf	pyg	pyh	65	pya	pyk	pyl	pym	pyn	pyo	
rav	raw	rae	raz	rab	rac	66	rad	raf	rag	rah	rai	rak	
rey	reu	rez	reb	rec	red	67	ref	reg	reh	rea	rek	rel	
rie	riz	rib	ric	rid	rif	68	rig	rih	ria	rik	ril	rim	
roz	rob	roc	rod	rof	rog	69	roh	roa	rok	rol	rom	ron	
rub	ruc	rud	ruf	rug	ruh	70	rua	ruk	rul	rum	run	ruo	
ryc	ryd	ryf	ryg	ryh	rya	71	ryk	ryl	rym	ryn	ryo	ryp	
saw	sae	saz	sab	sac	sad	72	saf	sag	sah	sai	sak	sal	
seu	sez	seb	sec	sed	sef	73	seg	seh	sea	sek	sel	sem	
siz	sib	sic	sid	sif	sig	74	sih	sia	sik	sil	sim	sin	
sob	soc	sod	sof	sog	soh	75	soa	sok	sol	som	son	soy	
suc	sud	suf	sug	suh	sua	76	suk	sul	sum	sun	suo	sup	
syd	syf	syg	syh	sya	syk	77	syl	sym	syn	syo	syp	syr	
tae	taz	tab	tac	tad	taf	78	tag	tah	tai	tak	tal	tam	
tez	teb	tec	ted	tef	teg	79	teh	tea	tek	tel	tem	ten	
tib	tic	tid	tif	tig	tih	80	tia	tik	til	tim	tin	tio	
toc	tod	tof	tog	toh	toa	81	tok	tol	tom	ton	toy	top	
tud	tuf	tug	tuh	tua	tuk	82	tul	tum	tun	tuo	tup	tur	
tyf	tyg	tyh	tya	tyk	tyl	83	tym	tyn	tyo	typ	tyr	tys	
vaz	vab	vac	vad	vaf	vag	84	vah	vai	vak	val	vam	van	
veb	vec	ved	vef	veg	veh	85	vea	vek	vel	vem	ven	veo	
vic	vid	vif	vig	vih	via	86	vik	vil	vim	vin	vio	vip	
vod	vof	vog	voh	voa	vok	87	vol	vom	von	voy	vop	vor	
vuf	vug	vuh	vua	vuk	vul	88	vum	vun	vuo	vup	vur	vus	
vyg	vyh	vya	vyk	vyl	vym	89	vyn	vyo	vyp	vyr	vys	vyt	
wab	wac	wad	waf	wag	wah	90	wai	wak	wal	wam	wan	wao	
wec	wed	wef	weg	weh	wea	91	wek	wel	wem	wen	weo	wep	
wid	wif	wig	wih	wia	wik	92	wil	wim	win	wio	wip	wir	
wof	wog	woh	woa	wok	wol	93	wom	won	woy	wop	wor	wos	
zad	zaf	zag	zah	zai	zak	94	zal	zam	zan	zao	zap	zar	
zef	zeg	zeh	zea	zek	zel	95	zem	zen	zeo	zep	zer	zes	
zig	zih	zia	zik	zil	zim	96	zin	zio	zip	zir	zis	zit	
zoh	zoa	zok	zol	zom	zon	97	zoy	zop	zor	zos	zot	zov	
zua	zuk	zul	zum	zun	zuo	98	zup	zur	zus	zut	zuv	zuw	
zyk	zyl	zym	zyn	zyo	zyp	99	zyr	zys	zyt	zyv	zyw	zye	

A

First 2 of the 5 letters: ac 30, et 31, ow 32, ux 33, ma 34, ve 35, li 36, ho 37, ru 38, xy 39

ac	et	ow	ux	ma	ve	li	ho	ru	xy
30	31	32	33	34	35	36	37	38	39

last 3 of the 5 letters:

A 30

1st	2nd	3rd	4th	5th	6th		7th	8th	9th	10th	11th	12th class
bag	bah	bai	bak	bal	bam	**00**	ban	bao	bap	bar	bas	bat
beh	bea	bek	bel	bem	ben	**01**	beo	bep	ber	bes	bet	bev
bia	bik	bil	bim	bin	bio	**02**	bip	bir	bis	bit	biv	biw
bok	bol	bom	bon	boy	bop	**03**	bor	bos	bot	bov	bow	boe
bul	bum	bun	buo	bup	bur	**04**	bus	but	buv	buw	bue	buz
bym	byn	byo	byp	byr	bys	**05**	byt	byv	byw	bye	byz	byb
cah	cai	cak	cal	cam	can	**06**	cao	cap	car	cas	cat	cav
cea	cek	cel	cem	cen	ceo	**07**	cep	cer	ces	cet	cev	cey
cik	cil	cim	cin	cio	cip	**08**	cir	cis	cit	civ	ciw	cie
col	com	con	coy	cop	cor	**09**	cos	cot	cov	cow	coe	coz
cum	cun	cuo	cup	cur	cus	**10**	cut	cuv	cuw	cue	cuz	cub
cyn	cyo	cyp	cyr	cys	cyt	**11**	cyv	cyw	cye	cyz	cyb	cyc
dai	dak	dal	dam	dan	dao	**12**	dap	dar	das	dat	dav	daw
dek	del	dem	den	deo	dep	**13**	der	des	det	dev	dey	deu
dil	dim	din	dio	dip	dir	**14**	dis	dit	div	diw	die	diz
dom	don	doy	dop	dor	dos	**15**	dot	dov	dow	doe	doz	dob
dun	duo	dup	dur	dus	dut	**16**	duv	duw	due	duz	dub	duc
dyo	dyp	dyr	dys	dyt	dyv	**17**	dyw	dye	dyz	dyb	dyc	dyd
fak	fal	fam	fan	fao	fap	**18**	far	fas	fat	fav	faw	fae
fel	fem	fen	feo	fep	fer	**19**	fes	fet	fev	fey	feu	fez
fim	fin	fio	fip	fir	fis	**20**	fit	fiv	fiw	fie	fiz	fib
fon	foy	fop	for	fos	fot	**21**	fov	fow	foe	foz	fob	foc
fuo	fup	fur	fus	fut	fuv	**22**	fuw	fue	fuz	fub	fuc	fud
fyp	fyr	fys	fyt	fyv	fyw	**23**	fye	fyz	fyb	fyc	fyd	fyf
gal	gam	gan	gao	gap	gar	**24**	gas	gat	gav	gaw	gae	gaz
gem	gen	geo	gep	ger	ges	**25**	get	gev	gey	geu	gez	geb
gin	gio	gip	gir	gis	git	**26**	giv	giw	gie	giz	gib	gic
goy	gop	gor	gos	got	gov	**27**	gow	goe	goz	gob	goc	god
gup	gur	gus	gut	guv	guw	**28**	gue	guz	gub	guc	gud	guf
gyr	gys	gyt	gyv	gyw	gye	**29**	gyz	gyb	gyc	gyd	gyf	gyg
ham	han	hao	hap	har	has	**30**	hat	hav	haw	hae	haz	hab
hen	heo	hep	her	hes	het	**31**	hev	hey	heu	hez	heb	hec
hio	hip	hir	his	hit	hiv	**32**	hiw	hie	hiz	hib	hic	hid
hop	hor	hos	hot	hov	how	**33**	hoe	hoz	hob	hoc	hod	hof
hur	hus	hut	huv	huw	hue	**34**	huz	hub	huc	hud	huf	hug
hys	hyt	hyv	hyw	hye	hyz	**35**	hyb	hyc	hyd	hyf	hyg	hyu
kao	kap	kar	kas	kat	kav	**36**	kaw	kae	kaz	kab	kac	kad
kep	ker	kes	ket	kev	key	**37**	keu	kez	keb	kec	ked	kef
kir	kis	kit	kiv	kiw	kie	**38**	kiz	kib	kic	kid	kif	kig
kos	kot	kov	kow	koe	koz	**39**	kob	koc	kod	kof	kog	koh
kut	kuv	kuw	kue	kuz	kub	**40**	kuc	kud	kuf	kug	kuh	kua
kyv	kyw	kye	kyz	kyb	kyc	**41**	kyd	kyf	kyg	kyh	kya	kyk
lap	lar	las	lat	lav	law	**42**	lae	laz	lab	lac	lad	laf
ler	les	let	lev	ley	leu	**43**	lez	leb	lec	led	lef	leg
lis	lit	liv	liw	lie	liz	**44**	lib	lic	lid	lif	lig	lih
lot	lov	low	loe	loz	lob	**45**	loc	lod	lof	log	loh	loa
luv	luw	lue	luz	lub	luc	**46**	lud	luf	lug	luh	lua	luk
lyw	lye	lyz	lyb	lyc	lyd	**47**	lyf	lyg	lyh	lya	lyk	lyl
mar	mas	mat	mav	maw	mae	**48**	maz	mab	mac	mad	maf	mag
mes	met	mev	mey	meu	mez	**49**	meb	mec	med	mef	meg	meh

A

First 2 of the 5 letters:	ac	et	ow	ux	ma	ve	li	ho	ru	xy
	30	31	32	33	34	35	36	37	38	39

last 3 of the 5 letters:

A
30

1st	2nd	3rd	4th	5th	6th		7th	8th	9th	10th	11th	12th class
mit	miv	miw	mie	miz	mib	**50**	mic	mid	mif	mig	mih	mia
mov	mow	moe	moz	mob	moc	**51**	mod	mof	mog	moh	moa	mok
muw	mue	muz	mub	muc	mud	**52**	muf	mug	muh	mua	muk	mul
mye	myz	myb	myc	myd	myf	**53**	myg	myh	mya	myk	myl	mym
nas	nat	nav	naw	nae	naz	**54**	nab	nac	nad	naf	nag	nah
net	nev	ney	neu	nez	neb	**55**	nec	ned	nef	neg	neh	nea
niv	niw	nie	niz	nib	nic	**56**	nid	nif	nig	nih	nia	nik
now	noe	noz	nob	noc	nod	**57**	nof	nog	noh	noa	nok	nol
nue	nuz	nub	nuc	nud	nuf	**58**	nug	nuh	nua	nuk	nul	num
nyz	nyb	nyc	nyd	nyf	nyg	**59**	nyh	nya	nyk	nyl	nym	nyn
pav	paw	pae	paz	pab	pac	**60**	pad	paf	pag	pah	pai	pak
pey	peu	pez	peb	pec	ped	**61**	pef	peg	peh	pea	pek	pel
pie	piz	pib	pic	pid	pif	**62**	pig	pih	pia	pik	pil	pim
poz	pob	poc	pod	pof	pog	**63**	poh	poa	pok	pol	pom	pon
pub	puc	pud	puf	pug	puh	**64**	pua	puk	pul	pum	pun	puo
pyc	pyd	pyf	pyg	pyh	pya	**65**	pyk	pyl	pym	pyn	pyo	pyp
raw	rae	raz	rab	rac	rad	**66**	raf	rag	rah	rai	rak	ral
reu	rez	reb	rec	red	ref	**67**	reg	reh	rea	rek	rel	rem
riz	rib	ric	rid	rif	rig	**68**	rih	ria	rik	ril	rim	rin
rob	roc	rod	rof	rog	roh	**69**	roa	rok	rol	rom	ron	roy
ruc	rud	ruf	rug	ruh	rua	**70**	ruk	rul	rum	run	ruo	rup
ryd	ryf	ryg	ryh	rya	ryk	**71**	ryl	rym	ryn	ryo	ryp	ryr
sae	saz	sab	sac	sad	saf	**72**	sag	sah	sai	sak	sal	sam
sez	seb	sec	sed	sef	seg	**73**	seh	sea	sek	sel	sem	sen
sib	sic	sid	sif	sig	sih	**74**	sia	sik	sil	sim	sin	sio
soc	sod	sof	sog	soh	soa	**75**	sok	sol	som	son	soy	sop
sud	suf	sug	suh	sua	suk	**76**	sul	sum	sun	suo	sup	sur
syf	syg	syh	sya	syk	syl	**77**	sym	syn	syo	syp	syr	sys
taz	tab	tac	tad	taf	tag	**78**	tah	tai	tak	tal	tam	tan
teb	tec	ted	tef	teg	teh	**79**	tea	tek	tel	tem	ten	teo
tic	tid	tif	tig	tih	tia	**80**	tik	til	tim	tin	tio	tip
tod	tof	tog	toh	toa	tok	**81**	tol	tom	ton	toy	top	tor
tuf	tug	tuh	tua	tuk	tul	**82**	tum	tun	tuo	tup	tur	tus
tyg	tyh	tya	tyk	tyl	tym	**83**	tyn	tyo	typ	tyr	tys	tyt
vab	vac	vad	vaf	vag	vah	**84**	vai	vak	val	vam	van	vao
vec	ved	vef	veg	veh	vea	**85**	vek	vel	vem	ven	veo	vep
vid	vif	vig	vih	via	vik	**86**	vil	vim	vin	vio	vip	vir
vof	vog	voh	voa	vok	vol	**87**	vom	von	voy	vop	vor	vos
vug	vuh	vua	vuk	vul	vum	**88**	vun	vuo	vup	vur	vus	vut
vyh	vya	vyk	vyl	vym	vyn	**89**	vyo	vyp	vyr	vys	vyt	vyv
wac	wad	waf	wag	wah	wai	**90**	wak	wal	wam	wan	wao	wap
wed	wef	weg	weh	wea	wek	**91**	wel	wem	wen	weo	wep	wer
wif	wig	wih	wia	wik	wil	**92**	wim	win	wio	wip	wir	wis
wog	woh	woa	wok	wol	wom	**93**	won	woy	wop	wor	wos	wot
zaf	zag	zah	zai	zak	zal	**94**	zam	zan	zao	zap	zar	zas
zeg	zeh	zea	zek	zel	zem	**95**	zen	zeo	zep	zer	zes	zet
zih	zia	zik	zil	zim	zin	**96**	zio	zip	zir	zis	zit	ziv
zoa	zok	zol	zom	zon	zoy	**97**	zop	zor	zos	zot	zov	zow
zuk	zul	zum	zun	zuo	zup	**98**	zur	zus	zut	zuv	zuw	zue
zyl	zym	zyn	zyo	zyp	zyr	**99**	zys	zyt	zyv	zyw	zye	zyz

A

First 2 of the 5 letters	ak	es	il	ov	ur	na	le	ji	so	pu
	40	41	42	43	44	45	46	47	48	49

last 3 of the 5 letters:

	1st	2nd	3rd	4th	5th	6th		7th	8th	9th	10th	11th	12th class
A	bah	bai	bak	bal	bam	ban	**00**	bao	bap	bar	bas	bat	bav
	bea	bek	bel	bem	ben	beo	**01**	bep	ber	bes	bet	bev	bey
40	bik	bil	bim	bin	bio	bip	**02**	bir	bis	bit	biv	biw	bie
	bol	bom	bon	boy	bop	bor	**03**	bos	bot	bov	bow	boe	boz
	bum	bun	buo	bup	bur	bus	**04**	but	buv	buw	bue	buz	bub
	byn	byo	byp	byr	bys	byt	**05**	byv	byw	bye	byz	byb	byc
	cai	cak	cal	cam	can	cao	**06**	cap	car	cas	cat	cav	caw
	cek	cel	cem	cen	ceo	cep	**07**	cer	ces	cet	cev	cey	ceu
	cil	cim	cin	cio	cip	cir	**08**	cis	cit	civ	ciw	cie	ciz
	com	con	coy	cop	cor	cos	**09**	cot	cov	cow	coe	coz	cob
	cun	cuo	cup	cur	cus	cut	**10**	cuv	cuw	cue	cuz	cub	cuc
	cyo	cyp	cyr	cys	cyt	cyv	**11**	cyw	cye	cyz	cyb	cyc	cyd
	dak	dal	dam	dan	dao	dap	**12**	dar	das	dat	dav	daw	dae
	del	dem	den	deo	dep	der	**13**	des	det	dev	dey	deu	dez
	dim	din	dio	dip	dir	dis	**14**	dit	div	diw	die	diz	dib
	don	doy	dop	dor	dos	dot	**15**	dov	dow	doe	doz	dob	doc
	duo	dup	dur	dus	dut	duv	**16**	duw	due	duz	dub	duc	dud
	dyp	dyr	dys	dyt	dyv	dyw	**17**	dye	dyz	dyb	dyc	dyd	dyf
	fal	fam	fan	fao	fap	far	**18**	fas	fat	fav	faw	fae	faz
	fem	fen	feo	fep	fer	fes	**19**	fet	fev	fey	feu	fez	feb
	fin	fio	fip	fir	fis	fit	**20**	fiv	fiw	fie	fiz	fib	fic
	foy	fop	for	fos	fot	fov	**21**	fow	foe	foz	fob	foc	fod
	fup	fur	fus	fut	fuv	fuw	**22**	fue	fuz	fub	fuc	fud	fuf
	fyr	fys	fyt	fyv	fyw	fye	**23**	fyz	fyb	fyc	fyd	fyf	fyg
	gam	gan	gao	gap	gar	gas	**24**	gat	gav	gaw	gae	gaz	gab
	gen	geo	gep	ger	ges	get	**25**	gev	gey	geu	gez	geb	gec
	gio	gip	gir	gis	git	giv	**26**	giw	gie	giz	gib	gic	gid
	gop	gor	gos	got	gov	gow	**27**	goe	goz	gob	goc	god	gof
	gur	gus	gut	guv	guw	gue	**28**	guz	gub	guc	gud	guf	gug
	gys	gyt	gyv	gyw	gye	gyz	**29**	gyb	gyc	gyd	gyf	gyg	gyh
	han	hao	hap	har	has	hat	**30**	hav	haw	hae	haz	hab	hac
	heo	hep	her	hes	het	hev	**31**	hey	heu	hez	heb	hec	hed
	hip	hir	his	hit	hiv	hiw	**32**	hie	hiz	hib	hic	hid	hif
	hor	hos	hot	hov	how	hoe	**33**	hoz	hob	hoc	hod	hof	hog
	hus	hut	huv	huw	hue	huz	**34**	hub	huc	hud	huf	hug	hui
	hyt	hyv	hyw	hye	hyz	hyb	**35**	hyc	hyd	hyf	hyg	hyu	hya
	kap	kar	kas	kat	kav	kaw	**36**	kae	kaz	kab	kac	kad	kaf
	ker	kes	ket	kev	key	keu	**37**	kez	keb	kec	ked	kef	keg
	kis	kit	kiv	kiw	kie	kiz	**38**	kib	kic	kid	kif	kig	kih
	kot	kov	kow	koe	koz	kob	**39**	koc	kod	kof	kog	koh	koa
	kuv	kuw	kue	kuz	kub	kuc	**40**	kud	kuf	kug	kuh	kua	kuk
	kyw	kye	kyz	kyb	kyc	kyd	**41**	kyf	kyg	kyh	kya	kyk	kyl
	lar	las	lat	lav	law	lae	**42**	laz	lab	lac	lad	laf	lag
	les	let	lev	ley	leu	lez	**43**	leb	lec	led	lef	leg	leh
	lit	liv	liw	lie	liz	lib	**44**	lic	lid	lif	lig	lih	lia
	lov	low	loe	loz	lob	loc	**45**	lod	lof	log	loh	loa	lok
	luw	lue	luz	lub	luc	lud	**46**	luf	lug	luh	lua	luk	lul
	lye	lyz	lyb	lyc	lyd	lyf	**47**	lyg	lyh	lya	lyk	lyl	lym
	mas	mat	mav	maw	mae	maz	**48**	mab	mac	mad	maf	mag	mah
	met	mev	mey	meu	mez	meb	**49**	mec	med	mef	meg	meh	mea

A

First 2 of the 5 letters	ak	es	il	ov	ur	na	le	ji	so	pu
	40	41	42	43	44	45	46	47	48	49

last 3 of the 5 letters:

1st	2nd	3rd	4th	5th	6th		7th	8th	9th	10th	11th	12th class	
miv	miw	mie	miz	mib	mic	**50**	mid	mif	mig	mih	mia	mik	**A**
mow	moe	moz	mob	moc	mod	**51**	mof	mog	moh	moa	mok	mol	
mue	muz	mub	muc	mud	muf	**52**	mug	muh	mua	muk	mul	mum	**40**
myz	myb	myc	myd	myf	myg	**53**	myh	mya	myk	myl	mym	myn	
nat	nav	naw	nae	naz	nab	**54**	nac	nad	naf	nag	nah	nai	
nev	ney	neu	nez	neb	nec	**55**	ned	nef	neg	neh	nea	nek	
niw	nie	niz	nib	nic	nid	**56**	nif	nig	nih	nia	nik	nil	
noe	noz	nob	noc	nod	nof	**57**	nog	noh	noa	nok	nol	nom	
nuz	nub	nuc	nud	nuf	nug	**58**	nuh	nua	nuk	nul	num	nun	
nyb	nyc	nyd	nyf	nyg	nyh	**59**	nya	nyk	nyl	nym	nyn	nyo	
paw	pae	paz	pab	pac	pad	**60**	paf	pag	pah	pai	pak	pal	
peu	pez	peb	pec	ped	pef	**61**	peg	peh	pea	pek	pel	pem	
piz	pib	pic	pid	pif	pig	**62**	pih	pia	pik	pil	pim	pin	
pob	poc	pod	pof	pog	poh	**63**	poa	pok	pol	pom	pon	poy	
puc	pud	puf	pug	puh	pua	**64**	puk	pul	pum	pun	puo	pup	
pyd	pyf	pyg	pyh	pya	pyk	**65**	pyl	pym	pyn	pyo	pyp	pyr	
rae	raz	rab	rac	rad	raf	**66**	rag	rah	rai	rak	ral	ram	
rez	reb	rec	red	ref	reg	**67**	reh	rea	rek	rel	rem	ren	
rib	ric	rid	rif	rig	rih	**68**	ria	rik	ril	rim	rin	rio	
roc	rod	rof	rog	roh	roa	**69**	rok	rol	rom	ron	roy	rop	
rud	ruf	rug	ruh	rua	ruk	**70**	rul	rum	run	ruo	rup	rur	
ryf	ryg	ryh	rya	ryk	ryl	**71**	rym	ryn	ryo	ryp	ryr	rys	
saz	sab	sac	sad	saf	sag	**72**	sah	sai	sak	sal	sam	san	
seb	sec	sed	sef	seg	seh	**73**	sea	sek	sel	sem	sen	seo	
sic	sid	sif	sig	sih	sia	**74**	sik	sil	sim	sin	sio	sip	
sod	sof	sog	soh	soa	sok	**75**	sol	som	son	soy	sop	sor	
suf	sug	suh	sua	suk	sul	**76**	sum	sun	suo	sup	sur	sus	
syg	syh	sya	syk	syl	sym	**77**	syn	syo	syp	syr	sys	syt	
tab	tac	tad	taf	tag	tah	**78**	tai	tak	tal	tam	tan	tao	
tec	ted	tef	teg	teh	tea	**79**	tek	tel	tem	ten	teo	tep	
tid	tif	tig	tih	tia	tik	**80**	til	tim	tin	tio	tip	tir	
tof	tog	toh	toa	tok	tol	**81**	tom	ton	toy	top	tor	tos	
tug	tuh	tua	tuk	tul	tum	**82**	tun	tuo	tup	tur	tus	tut	
tyh	tya	tyk	tyl	tym	tyn	**83**	tyo	typ	tyr	tys	tyt	tyv	
vac	vad	vaf	vag	vah	vai	**84**	vak	val	vam	van	vao	vap	
ved	vef	veg	veh	vea	vek	**85**	vel	vem	ven	veo	vep	ver	
vif	vig	vih	via	vik	vil	**86**	vim	vin	vio	vip	vir	vis	
vog	voh	voa	vok	vol	vom	**87**	von	voy	vop	vor	vos	vot	
vuh	vua	vuk	vul	vum	vun	**88**	vuo	vup	vur	vus	vut	vuv	
vya	vyk	vyl	vym	vyn	vyo	**89**	vyp	vyr	vys	vyt	vyv	vyw	
wad	waf	wag	wah	wai	wak	**90**	wal	wam	wan	wao	wap	war	
wef	weg	weh	wea	wek	wel	**91**	wem	wen	weo	wep	wer	wes	
wig	wih	wia	wik	wil	wim	**92**	win	wio	wip	wir	wis	wit	
woh	woa	wok	wol	wom	won	**93**	woy	wop	wor	wos	wot	wov	
zag	zah	zai	zak	zal	zam	**94**	zan	zao	zap	zar	zas	zat	
zeh	zea	zek	zel	zem	zen	**95**	zeo	zep	zer	zes	zet	zev	
zia	zik	zil	zim	zin	zio	**96**	zip	zir	zis	zit	ziv	ziw	
zok	zol	zom	zon	zoy	zop	**97**	zor	zos	zot	zov	zow	zoe	
zul	zum	zun	zuo	zup	zur	**98**	zus	zut	zuv	zuw	zue	zuz	
zym	zyn	zyo	zyp	zyr	zys	**99**	zyt	zyv	zyw	zye	zyz	zyb	

A

First 2 of the 5 letters:	ex	id	of	uz	pa	ne	ki	jo	fu	gy
last 3 of the 5 letters:	50	51	52	53	54	55	56	57	58	59

	1st	2nd	3rd	4th	5th	6th		7th	8th	9th	10th	11th	12th class
A	bai	bak	bal	bam	ban	bao	**00**	bap	bar	bas	bat	bav	baw
	bek	bel	bem	ben	beo	bep	**01**	ber	bes	bet	bev	bey	beu
50	bil	bim	bin	bio	bip	bir	**02**	bis	bit	biv	biw	bie	biz
	bom	bon	boy	bop	bor	bos	**03**	bot	bov	bow	boe	boz	bob
	bun	buo	bup	bur	bus	but	**04**	buv	buw	bue	buz	bub	buc
	byo	byp	byr	bys	byt	byv	**05**	byw	bye	byz	byb	byc	byd
	cak	cal	cam	can	cao	cap	**06**	car	cas	cat	cav	caw	cae
	cel	cem	cen	ceo	cep	cer	**07**	ces	cet	cev	cey	ceu	cez
	cim	cin	cio	cip	cir	cis	**08**	cit	civ	ciw	cie	ciz	cib
	con	coy	cop	cor	cos	cot	**09**	cov	cow	coe	coz	cob	coc
	cuo	cup	cur	cus	cut	cuv	**10**	cuw	cue	cuz	cub	cuc	cud
	cyp	cyr	cys	cyt	cyv	cyw	**11**	cye	cyz	cyb	cyc	cyd	cyf
	dal	dam	dan	dao	dap	dar	**12**	das	dat	dav	daw	dae	daz
	dem	den	deo	dep	der	des	**13**	det	dev	dey	deu	dez	deb
	din	dio	dip	dir	dis	dit	**14**	div	diw	die	diz	dib	dic
	doy	dop	dor	dos	dot	dov	**15**	dow	doe	doz	dob	doc	dod
	dup	dur	dus	dut	duv	duw	**16**	due	duz	dub	duc	dud	duf
	dyr	dys	dyt	dyv	dyw	dye	**17**	dyz	dyb	dyc	dyd	dyf	dyg
	fam	fan	fao	fap	far	fas	**18**	fat	fav	faw	fae	faz	fab
	fen	feo	fep	fer	fes	fet	**19**	fev	fey	feu	fez	feb	fec
	fio	fip	fir	fis	fit	fiv	**20**	fiw	fie	fiz	fib	fic	fid
	fop	for	fos	fot	fov	fow	**21**	foe	foz	fob	foc	fod	fof
	fur	fus	fut	fuv	fuw	fue	**22**	fuz	fub	fuc	fud	fuf	fug
	fys	fyt	fyv	fyw	fye	fyz	**23**	fyb	fyc	fyd	fyf	fyg	fyh
	gan	gao	gap	gar	gas	gat	**24**	gav	gaw	gae	gaz	gab	gac
	geo	gep	ger	ges	get	gev	**25**	gey	geu	gez	geb	gec	ged
	gip	gir	gis	git	giv	giw	**26**	gie	giz	gib	gic	gid	gif
	gor	gos	got	gov	gow	goe	**27**	goz	gob	goc	god	gof	gog
	gus	gut	guv	guw	gue	guz	**28**	gub	guc	gud	guf	gug	guh
	gyt	gyv	gyw	gye	gyz	gyb	**29**	gyc	gyd	gyf	gyg	gyh	gya
	hao	hap	har	has	hat	hav	**30**	haw	hae	haz	hab	hac	had
	hep	her	hes	het	hev	hey	**31**	heu	hez	heb	hec	hed	hef
	hir	his	hit	hiv	hiw	hie	**32**	hiz	hib	hic	hid	hif	hig
	hos	hot	hov	how	hoe	hoz	**33**	hob	hoc	hod	hof	hog	hoi
	hut	huv	huw	hue	huz	hub	**34**	huc	hud	huf	hug	hui	hua
	hyv	hyw	hye	hyz	hyb	hyc	**35**	hyd	hyf	hyg	hyu	hya	hyk
	kar	kas	kat	kav	kaw	kae	**36**	kaz	kab	kac	kad	kaf	kag
	kes	ket	kev	key	keu	kez	**37**	keb	kec	ked	kef	keg	keh
	kit	kiv	kiw	kie	kiz	kib	**38**	kic	kid	kif	kig	kih	kia
	kov	kow	koe	koz	kob	koc	**39**	kod	kof	kog	koh	koa	kok
	kuw	kue	kuz	kub	kuc	kud	**40**	kuf	kug	kuh	kua	kuk	kul
	kye	kyz	kyb	kyc	kyd	kyf	**41**	kyg	kyh	kya	kyk	kyl	kym
	las	lat	lav	law	lae	laz	**42**	lab	lac	lad	laf	lag	lah
	let	lev	ley	leu	lez	leb	**43**	lec	led	lef	leg	leh	lea
	liv	liw	lie	liz	lib	lic	**44**	lid	lif	lig	lih	lia	lik
	low	loe	loz	lob	loc	lod	**45**	lof	log	loh	loa	lok	lol
	lue	luz	lub	luc	lud	luf	**46**	lug	luh	lua	luk	lul	lum
	lyz	lyb	lyc	lyd	lyf	lyg	**47**	lyh	lya	lyk	lyl	lym	lyn
	mat	mav	maw	mae	maz	mab	**48**	mac	mad	maf	mag	mah	mai

A

First 2 of the 5 letters:	ex	id	of	uz	pa	ne	ki	jo	fu	gy
	50	51	52	53	54	55	56	57	58	59

last 3 of the 5 letters:

1st	2nd	3rd	4th	5th	6th		7th	8th	9th	10th	11th	12th class
miw	mie	miz	mib	mic	mid	50	mif	mig	mih	mia	mik	mil
moe	moz	mob	moc	mod	mof	51	mog	moh	moa	mok	mol	mom
muz	mub	muc	mud	muf	mug	52	muh	mua	muk	mul	mum	mun
myb	myc	myd	myf	myg	myh	53	mya	myk	myl	mym	myn	myo
nav	naw	nae	naz	nab	nac	54	nad	naf	nag	nah	nai	nak
ney	neu	nez	neb	nec	ned	55	nef	neg	neh	nea	nek	nel
nie	niz	nib	nic	nid	nif	56	nig	nih	nia	nik	nil	nim
noz	nob	noc	nod	nof	nog	57	noh	noa	nok	nol	nom	non
nub	nuc	nud	nuf	nug	nuh	58	nua	nuk	nul	num	nun	nuo
nyc	nyd	nyf	nyg	nyh	nya	59	nyk	nyl	nym	nyn	nyo	nyp
pae	paz	pab	pac	pad	paf	60	pag	pah	pai	pak	pal	pam
pez	peb	pec	ped	pef	peg	61	peh	pea	pek	pel	pem	pen
pib	pic	pid	pif	pig	pih	62	pia	pik	pil	pim	pin	pio
poc	pod	pof	pog	poh	poa	63	pok	pol	pom	pon	poy	pop
pud	puf	pug	puh	pua	puk	64	pul	pum	pun	puo	pup	pur
pyf	pyg	pyh	pya	pyk	pyl	65	pym	pyn	pyo	pyp	pyr	pys
raz	rab	rac	rad	raf	rag	66	rah	rai	rak	ral	ram	ran
reb	rec	red	ref	reg	reh	67	rea	rek	rel	rem	ren	reo
ric	rid	rif	rig	rih	ria	68	rik	ril	rim	rin	rio	rip
rod	rof	rog	roh	roa	rok	69	rol	rom	ron	roy	rop	ror
ruf	rug	ruh	rua	ruk	rul	70	rum	run	ruo	rup	rur	rus
ryg	ryh	rya	ryk	ryl	rym	71	ryn	ryo	ryp	ryr	rys	ryt
sab	sac	sad	saf	sag	sah	72	sai	sak	sal	sam	san	sao
sec	sed	sef	seg	seh	sea	73	sek	sel	sem	sen	seo	sep
sid	sif	sig	sih	sia	sik	74	sil	sim	sin	sio	sip	sir
sof	sog	soh	soa	sok	sol	75	som	son	soy	sop	sor	sos
sug	suh	sua	suk	sul	sum	76	sun	suo	sup	sur	sus	sut
syh	sya	syk	syl	sym	syn	77	syo	syp	syr	sys	syt	syv
tac	tad	taf	tag	tah	tai	78	tak	tal	tam	tan	tao	tap
ted	tef	teg	teh	tea	tek	79	tel	tem	ten	teo	tep	ter
tif	tig	tih	tia	tik	til	80	tim	tin	tio	tip	tir	tis
tog	toh	toa	tok	tol	tom	81	ton	toy	top	tor	tos	tot
tuh	tua	tuk	tul	tum	tun	82	tuo	tup	tur	tus	tut	tuv
tya	tyk	tyl	tym	tyn	tyo	83	typ	tyr	tys	tyt	tyv	tyw
vad	vaf	vag	vah	vai	vak	84	val	vam	van	vao	vap	var
vef	veg	veh	vea	vek	vel	85	vem	ven	veo	vep	ver	ves
vig	vih	via	vik	vil	vim	86	vin	vio	vip	vir	vis	vit
voh	voa	vok	vol	vom	von	87	voy	vop	vor	vos	vot	vov
vua	vuk	vul	vum	vun	vuo	88	vup	vur	vus	vut	vuv	vuw
vyk	vyl	vym	vyn	vyo	vyp	89	vyr	vys	vyt	vyv	vyw	vye
waf	wag	wah	wai	wak	wal	90	wam	wan	wao	wap	war	was
weg	weh	wea	wek	wel	wem	91	wen	weo	wep	wer	wes	wet
wih	wia	wik	wil	wim	win	92	wio	wip	wir	wis	wit	wiv
woa	wok	wol	wom	won	woy	93	wop	wor	wos	wot	wov	wue
zah	zai	zak	zal	zam	zan	94	zao	zap	zar	zas	zat	zav
zea	zek	zel	zem	zen	zeo	95	zep	zer	zes	zet	zev	zey
zik	zil	zim	zin	zio	zip	96	zir	zis	zit	ziv	ziw	zie
zol	zom	zon	zoy	zop	zor	97	zos	zot	zov	zow	zoe	zoz
zum	zun	zuo	zup	zur	zus	98	zut	zuv	zuw	zue	zuz	zub
zyn	zyo	zyp	zyr	zys	zyt	99	zyv	zyw	zye	zyz	zyb	zyc

A 50

A

First 2 of the 5 letters:	aj	it	ox	uf	ra	se	zi	lo	xu	by
	60	61	62	63	64	65	66	67	68	69

last 3 of the 5 letters:

A 60

1st	2nd	3rd	4th	5th	6th		7th	8th	9th	10th	11th	12th class
bak	bal	bam	ban	bao	bap	**00**	bar	bas	bat	bav	baw	bae
bel	bem	ben	beo	bep	ber	**01**	bes	bet	bev	bey	beu	bez
bim	bin	bio	bip	bir	bis	**02**	bit	biv	biw	bie	biz	bib
bon	boy	bop	bor	bos	bot	**03**	bov	bow	boe	boz	bob	boc
buo	bup	bur	bus	but	buv	**04**	buw	bue	buz	bub	buc	bud
byp	byr	bys	byt	byv	byw	**05**	bye	byz	byb	byc	byd	byf
cal	cam	can	cao	cap	car	**06**	cas	cat	cav	caw	cae	caz
cem	cen	ceo	cep	cer	ces	**07**	cet	cev	cey	ceu	cez	ceb
cin	cio	cip	cir	cis	cit	**08**	civ	ciw	cie	ciz	cib	cic
coy	cop	cor	cos	cot	cov	**09**	cow	coe	coz	cob	coc	cod
cup	cur	cus	cut	cuv	cuw	**10**	cue	cuz	cub	cuc	cud	cuf
cyr	cys	cyt	cyv	cyw	cye	**11**	cyz	cyb	cyc	cyd	cyf	cyg
dam	dan	dao	dap	dar	das	**12**	dat	dav	daw	dae	daz	dab
den	deo	dep	der	des	det	**13**	dev	dey	deu	dez	deb	dec
dio	dip	dir	dis	dit	div	**14**	diw	die	diz	dib	dic	did
dop	dor	dos	dot	dov	dow	**15**	doe	doz	dob	doc	dod	dof
dur	dus	dut	duv	duw	due	**16**	duz	dub	duc	dud	duf	dug
dys	dyt	dyv	dyw	dye	dyz	**17**	dyb	dyc	dyd	dyf	dyg	dyh
fan	fao	fap	far	fas	fat	**18**	fav	faw	fae	faz	fab	fac
feo	fep	fer	fes	fet	fev	**19**	fey	feu	fez	feb	fec	fed
fip	fir	fis	fit	fiv	fiw	**20**	fie	fiz	fib	fic	fid	fif
for	fos	fot	fov	fow	foe	**21**	foz	fob	foc	fod	fof	fog
fus	fut	fuv	fuw	fue	fuz	**22**	fub	fuc	fud	fuf	fug	fuh
fyt	fyv	fyw	fye	fyz	fyb	**23**	fyc	fyd	fyf	fyg	fyh	fya
gao	gap	gar	gas	gat	gav	**24**	gaw	gae	gaz	gab	gac	gad
gep	ger	ges	get	gev	gey	**25**	geu	gez	geb	gec	ged	gef
gir	gis	git	giv	giw	gie	**26**	giz	gib	gic	gid	gif	gig
gos	got	gov	gow	goe	goz	**27**	gob	goc	god	gof	gog	goh
gut	guv	guw	gue	guz	gub	**28**	guc	gud	guf	gug	guh	gua
gyv	gyw	gye	gyz	gyb	gyc	**29**	gyd	gyf	gyg	gyh	gya	gyk
hap	har	has	hat	hav	haw	**30**	hae	haz	hab	hac	had	haf
her	hes	het	hev	hey	heu	**31**	hez	heb	hec	hed	hef	heg
his	hit	hiv	hiw	hie	hiz	**32**	hib	hic	hid	hif	hig	hiu
hot	hov	how	hoe	hoz	hob	**33**	hoc	hod	hof	hog	hoi	hoa
huv	huw	hue	huz	hub	huc	**34**	hud	huf	hug	hui	hua	huk
hyw	hye	hyz	hyb	hyc	hyd	**35**	hyf	hyg	hyu	hya	hyk	hyl
kas	kat	kav	kaw	kae	kaz	**36**	kab	kac	kad	kaf	kag	kah
ket	kev	key	keu	kez	keb	**37**	kec	ked	kef	keg	keh	kea
kiv	kiw	kie	kiz	kib	kic	**38**	kid	kif	kig	kih	kia	kik
kow	koe	koz	kob	koc	kod	**39**	kof	kog	koh	koa	kok	kol
kue	kuz	kub	kuc	kud	kuf	**40**	kug	kuh	kua	kuk	kul	kum
kyz	kyb	kyc	kyd	kyf	kyg	**41**	kyh	kya	kyk	kyl	kym	kyn
lat	lav	law	lae	laz	lab	**42**	lac	lad	laf	lag	lah	lai
lev	ley	leu	lez	leb	lec	**43**	led	lef	leg	leh	lea	lek
liw	lie	liz	lib	lic	lid	**44**	lif	lig	lih	lia	lik	lil
loe	loz	lob	loc	lod	lof	**45**	log	loh	loa	lok	lol	lom
luz	lub	luc	lud	luf	lug	**46**	luh	lua	luk	lul	lum	lun
lyb	lyc	lyd	lyf	lyg	lyh	**47**	lya	lyk	lyl	lym	lyn	lyo
mav	maw	mae	maz	mab	mac	**48**	mad	maf	mag	mah	mai	mak
mey	meu	mez	meb	mec	med	**49**	mef	meg	meh	mea	mek	mel

A

First 2 of the 5 letters:

aj	it	ox	uf	ra	se	zi	lo	xu	by
60	61	62	63	64	65	66	67	68	69

last 3 of the 5 letters:

A 60

1st	2nd	3rd	4th	5th	6th		7th	8th	9th	10th	11th	12th class
mie	miz	mib	mic	mid	mif	50	mig	mih	mia	mik	mil	mim
moz	mob	moc	mod	mof	mog	51	moh	moa	mok	mol	mom	mon
mub	muc	mud	muf	mug	muh	52	mua	muk	mul	mum	mun	muo
myc	myd	myf	myg	myh	mya	53	myk	myl	mym	myn	myo	myp
naw	nae	naz	nab	nac	nad	54	naf	nag	nah	nai	nak	nal
neu	nez	neb	nec	ned	nef	55	neg	neh	nea	nek	nel	nem
niz	nib	nic	nid	nif	nig	56	nih	nia	nik	nil	nim	nin
nob	noc	nod	nof	nog	noh	57	noa	nok	nol	nom	non	noy
nuc	nud	nuf	nug	nuh	nua	58	nuk	nul	num	nun	nuo	nup
nyd	nyf	nyg	nyh	nya	nyk	59	nyl	nym	nyn	nyo	nyp	nyr
paz	pab	pac	pad	paf	pag	60	pah	pai	pak	pal	pam	pan
peb	pec	ped	pef	peg	peh	61	pea	pek	pel	pem	pen	peo
pic	pid	pif	pig	pih	pia	62	pik	pil	pim	pin	pio	pip
pod	pof	pog	poh	poa	pok	63	pol	pom	pon	poy	pop	por
puf	pug	puh	pua	puk	pul	64	pum	pun	puo	pup	pur	pus
pyg	pyh	pya	pyk	pyl	pym	65	pyn	pyo	pyp	pyr	pys	pyt
rab	rac	rad	raf	rag	rah	66	rai	rak	ral	ram	ran	rao
rec	red	ref	reg	reh	rea	67	rek	rel	rem	ren	reo	rep
rid	rif	rig	rih	ria	rik	68	ril	rim	rin	rio	rip	rir
rof	rog	roh	roa	rok	rol	69	rom	ron	roy	rop	ror	ros
rug	ruh	rua	ruk	rul	rum	70	run	ruo	rup	rur	rus	rut
ryh	rya	ryk	ryl	rym	ryn	71	ryo	ryp	ryr	rys	ryt	ryv
sac	sad	saf	sag	sah	sai	72	sak	sal	sam	san	sao	sap
sed	sef	seg	seh	sea	sek	73	sel	sem	sen	seo	sep	ser
sif	sig	sih	sia	sik	sil	74	sim	sin	sio	sip	sir	sis
sog	soh	soa	sok	sol	som	75	son	soy	sop	sor	sos	sot
suh	sua	suk	sul	sum	sun	76	suo	sup	sur	sus	sut	suv
sya	syk	syl	sym	syn	syo	77	syp	syr	sys	syt	syv	syw
tad	taf	tag	tah	tai	tak	78	tal	tam	tan	tao	tap	tar
tef	teg	teh	tea	tek	tel	79	tem	ten	teo	tep	ter	tes
tig	tih	tia	tik	til	tim	80	tin	tio	tip	tir	tis	tit
toh	toa	tok	tol	tom	ton	81	toy	top	tor	tos	tot	tov
tua	tuk	tul	tum	tun	tuo	82	tup	tur	tus	tut	tuv	tuw
tyk	tyl	tym	tyn	tyo	typ	83	tyr	tys	tyt	tyv	tyw	tye
vaf	vag	vah	vai	vak	val	84	vam	van	vao	vap	var	vas
veg	veh	vea	vek	vel	vem	85	ven	veo	vep	ver	ves	vet
vih	via	vik	vil	vim	vin	86	vio	vip	vir	vis	vit	viv
voa	vok	vol	vom	von	voy	87	vop	vor	vos	vot	vov	vow
vuk	vul	vum	vun	vuo	vup	88	vur	vus	vut	vuv	vuw	vue
vyl	vym	vyn	vyo	vyp	vyr	89	vys	vyt	vyv	vyw	vye	vyz
wag	wah	wai	wak	wal	wam	90	wan	wao	wap	war	was	wat
weh	wea	wek	wel	wem	wen	91	weo	wep	wer	wes	wet	wev
wia	wik	wil	wim	win	wio	92	wip	wir	wis	wit	wiv	woe
wok	wol	wom	won	woy	wop	93	wor	wos	wot	wov	wue	wuz
zai	zak	zal	zam	zan	zao	94	zap	zar	zas	zat	zav	zaw
zek	zel	zem	zen	zeo	zep	95	zer	zes	zet	zev	zey	zeu
zil	zim	zin	zio	zip	zir	96	zis	zit	ziv	ziw	zie	ziz
zom	zon	zoy	zop	zor	zos	97	zot	zov	zow	zoe	zoz	zob
zun	zuo	zup	zur	zus	zut	98	zuv	zuw	zue	zuz	zub	zuc
zyo	zyp	zyr	zys	zyt	zyv	99	zyw	zye	zyz	zyb	zyc	zyd

A

First 2 of the 5 letters:	al	eb	im	ot	ug	sa	fe	bi	ku	ry
	70	71	72	73	74	75	76	77	78	79

last 3 of the 5 letters:

	1st	2nd	3rd	4th	5th	6th		7th	8th	9th	10th	11th	12th class
A	bal	bam	ban	bao	bap	bar	00	bas	bat	bav	baw	bae	baz
	bem	ben	beo	bep	ber	bes	01	bet	bev	bey	beu	bez	beb
70	bin	bio	bip	bir	bis	bit	02	biv	biw	bie	biz	bib	bic
	boy	bop	bor	bos	bot	bov	03	bow	boe	boz	bob	boc	bod
	bup	bur	bus	but	buv	buw	04	bue	buz	bub	buc	bud	buf
	byr	bys	byt	byv	byw	bye	05	byz	byb	byc	byd	byf	byg
	cam	can	cao	cap	car	cas	06	cat	cav	caw	cae	caz	cab
	cen	ceo	cep	cer	ces	cet	07	cev	cey	ceu	cez	ceb	cec
	cio	cip	cir	cis	cit	civ	08	ciw	cie	ciz	cib	cic	cid
	cop	cor	cos	cot	cov	cow	09	coe	coz	cob	coc	cod	cof
	cur	cus	cut	cuv	cuw	cue	10	cuz	cub	cuc	cud	cuf	cug
	cys	cyt	cyv	cyw	cye	cyz	11	cyb	cyc	cyd	cyf	cyg	cyh
	dan	dao	dap	dar	das	dat	12	dav	daw	dae	daz	dab	dac
	deo	dep	der	des	det	dev	13	dey	deu	dez	deb	dec	ded
	dip	dir	dis	dit	div	diw	14	die	diz	dib	dic	did	dif
	dor	dos	dot	dov	dow	doe	15	doz	dob	doc	dod	dof	dog
	dus	dut	duv	duw	due	duz	16	dub	duc	dud	duf	dug	duh
	dyt	dyv	dyw	dye	dyz	dyb	17	dyc	dyd	dyf	dyg	dyh	dya
	fao	fap	far	fas	fat	fav	18	faw	fae	faz	fab	fac	fad
	fep	fer	fes	fet	fev	fey	19	feu	fez	feb	fec	fed	fef
	fir	fis	fit	fiv	fiw	fie	20	fiz	fib	fic	fid	fif	fig
	fos	fot	fov	fow	foe	foz	21	fob	foc	fod	fof	fog	foh
	fut	fuv	fuw	fue	fuz	fub	22	fuc	fud	fuf	fug	fuh	fua
	fyv	fyw	fye	fyz	fyb	fyc	23	fyd	fyf	fyg	fyh	fya	fyk
	gap	gar	gas	gat	gav	gaw	24	gae	gaz	gab	gac	gad	gaf
	ger	ges	get	gev	gey	geu	25	gez	geb	gec	ged	gef	geg
	gis	git	giv	giw	gie	giz	26	gib	gic	gid	gif	gig	gih
	got	gov	gow	goe	goz	gob	27	goc	god	gof	gog	goh	goa
	guv	guw	gue	guz	gub	guc	28	gud	guf	gug	guh	gua	guk
	gyw	gye	gyz	gyb	gyc	gyd	29	gyf	gyg	gyh	gya	gyk	gyl
	har	has	hat	hav	haw	hae	30	haz	hab	hac	had	haf	hag
	hes	het	hev	hey	heu	hez	31	heb	hec	hed	hef	heg	hei
	hit	hiv	hiw	hie	hiz	hib	32	hic	hid	hif	hig	hiu	hia
	hov	how	hoe	hoz	hob	hoc	33	hod	hof	hog	hoi	hoa	hok
	huw	hue	huz	hub	huc	hud	34	huf	hug	hui	hua	huk	hul
	hye	hyz	hyb	hyc	hyd	hyf	35	hyg	hyu	hya	hyk	hyl	hym
	kat	kav	kaw	kae	kaz	kab	36	kac	kad	kaf	kag	kah	kai
	kev	key	keu	kez	keb	kec	37	ked	kef	keg	keh	kea	kek
	kiw	kie	kiz	kib	kic	kid	38	kif	kig	kih	kia	kik	kil
	koe	koz	kob	koc	kod	kof	39	kog	koh	koa	kok	kol	kom
	kuz	kub	kuc	kud	kuf	kug	40	kuh	kua	kuk	kul	kum	kun
	kyb	kyc	kyd	kyf	kyg	kyh	41	kya	kyk	kyl	kym	kyn	kyo
	lav	law	lae	laz	lab	lac	42	lad	laf	lag	lah	lai	lak
	ley	leu	lez	leb	lec	led	43	lef	leg	leh	lea	lek	lel
	lie	liz	lib	lic	lid	lif	44	lig	lih	lia	lik	lil	lim
	loz	lob	loc	lod	lof	log	45	loh	loa	lok	lol	lom	lon
	lub	luc	lud	luf	lug	luh	46	lua	luk	lul	lum	lun	luo
	lyc	lyd	lyf	lyg	lyh	lya	47	lyk	lyl	lym	lyn	lyo	lyp
	maw	mae	maz	mab	mac	mad	48	maf	mag	mah	mai	mak	mal
	meu	mez	meb	mec	med	mef	49	meg	meh	mea	mek	mel	mem

A

First 2 of the 5 letters:	al	eb	im	ot	ug	sa	fe	bi	ku	ry
	70	71	72	73	74	75	76	77	78	79

last 3 of the 5 letters:

1st	2nd	3rd	4th	5th	6th		7th	8th	9th	10th	11th	12th class
miz	mib	mic	mid	mif	mig	**50**	mih	mia	mik	mil	mim	min
mob	moc	mod	mof	mog	moh	**51**	moa	mok	mol	mom	mon	moy
muc	mud	muf	mug	muh	mua	**52**	muk	mul	mum	mun	muo	mup
myd	myf	myg	myh	mya	myk	**53**	myl	mym	myn	myo	myp	myr
nae	naz	nab	nac	nad	naf	**54**	nag	nah	nai	nak	nal	nam
nez	neb	nec	ned	nef	neg	**55**	neh	nea	nek	nel	nem	nen
nib	nic	nid	nif	nig	nih	**56**	nia	nik	nil	nim	nin	nio
noc	nod	nof	nog	noh	noa	**57**	nok	nol	nom	non	noy	nop
nud	nuf	nug	nuh	nua	nuk	**58**	nul	num	nun	nuo	nup	nur
nyf	nyg	nyh	nya	nyk	nyl	**59**	nym	nyn	nyo	nyp	nyr	nys
pab	pac	pad	paf	pag	pah	**60**	pai	pak	pal	pam	pan	pao
pec	ped	pef	peg	peh	pea	**61**	pek	pel	pem	pen	peo	pep
pid	pif	pig	pih	pia	pik	**62**	pil	pim	pin	pio	pip	pir
pof	pog	poh	poa	pok	pol	**63**	pom	pon	poy	pop	por	pos
pug	puh	pua	puk	pul	pum	**64**	pun	puo	pup	pur	pus	put
pyh	pya	pyk	pyl	pym	pyn	**65**	pyo	pyp	pyr	pys	pyt	pyv
rac	rad	raf	rag	rah	rai	**66**	rak	ral	ram	ran	rao	rap
red	ref	reg	reh	rea	rek	**67**	rel	rem	ren	reo	rep	rer
rif	rig	rih	ria	rik	ril	**68**	rim	rin	rio	rip	rir,	ris
rog	roh	roa	rok	rol	rom	**69**	ron	roy	rop	ror	ros	rot
ruh	rua	ruk	rul	rum	run	**70**	ruo	rup	rur	rus	rut	ruv
rya	ryk	ryl	rym	ryn	ryo	**71**	ryp	ryr	rys	ryt	ryv	ryw
sad	saf	sag	sah	sai	sak	**72**	sal	sam	san	sao	sap	sar
sef	seg	seh	sea	sek	sel	**73**	sem	sen	seo	sep	ser	ses
sig	sih	sia	sik	sil	sim	**74**	sin	sio	sip	sir	sis	sit
soh	soa	sok	sol	som	son	**75**	soy	sop	sor	sos	sot	sov
sua	suk	sul	sum	sun	suo	**76**	sup	sur	sus	sut	suv	suw
syk	syl	sym	syn	syo	syp	**77**	syr	sys	syt	syv	syw	sye
taf	tag	tah	tai	tak	tal	**78**	tam	tan	tao	tap	tar	tas
teg	teh	tea	tek	tel	tem	**79**	ten	teo	tep	ter	tes	tet
tih	tia	tik	til	tim	tin	**80**	tio	tip	tir	tis	tit	tiv
toa	tok	tol	tom	ton	toy	**81**	top	tor	tos	tot	tov	tow
tuk	tul	tum	tun	tuo	tup	**82**	tur	tus	tut	tuv	tuw	tue
tyl	tym	tyn	tyo	typ	tyr	**83**	tys	tyt	tyv	tyw	tye	tyz
vag	vah	vai	vak	val	vam	**84**	van	vao	vap	var	vas	vat
veh	vea	vek	vel	vem	ven	**85**	veo	vep	ver	ves	vet	vev
via	vik	vil	vim	vin	vio	**86**	vip	vir	vis	vit	viv	viw
vok	vol	vom	von	voy	vop	**87**	vor	vos	vot	vov	vow	voe
vul	vum	vun	vuo	vup	vur	**88**	vus	vut	vuv	vuw	vue	vuz
vym	vyn	vyo	vyp	vyr	vys	**89**	vyt	vyv	vyw	vye	vyz	vyb
wah	wai	wak	wal	wam	wan	**90**	wao	wap	war	was	wat	wav
wea	wek	wel	wem	wen	weo	**91**	wep	wer	wes	wet	wev	wey
wik	wil	wim	win	wio	wip	**92**	wir	wis	wit	wiv	woe	wie
wol	wom	won	woy	wop	wor	**93**	wos	wot	wov	wue	wuz	woz
zak	zal	zam	zan	zao	zap	**94**	zar	zas	zat	zav	zaw	zae
zel	zem	zen	zeo	zep	zer	**95**	zes	zet	zev	zey	zeu	zez
zim	zin	zio	zip	zir	zis	**96**	zit	ziv	ziw	zie	ziz	zib
zon	zoy	zop	zor	zos	zot	**97**	zov	zow	zoe	zoz	zob	zoc
zuo	zup	zur	zus	zut	zuv	**98**	zuw	zue	zuz	zub	zuc	zud
zyp	zyr	zys	zyt	zyv	zyw	**99**	zye	zyz	zyb	zyc	zyd	zyf

A

First 2 of the 5 letters:	ap	el	in	or	ut	ta	re	xo	vu	dy
last 3 of the 5 letters:	80	81	82	83	84	85	86	87	88	89

A 80

1st	2nd	3rd	4th	5th	6th		7th	8th	9th	10th	11th	12th class
bam	ban	bao	bap	bar	bas	**00**	bat	bav	baw	bae	baz	bab
ben	beo	bep	ber	bes	bet	**01**	bev	bey	beu	bez	beb	bec
bio	bip	bir	bis	bit	biv	**02**	biw	bie	biz	bib	bic	bid
bop	bor	bos	bot	bov	bow	**03**	boe	boz	bob	boc	bod	bof
bur	bus	but	buv	buw	bue	**04**	buz	bub	buc	bud	buf	bug
bys	byt	byv	byw	bye	byz	**05**	byb	byc	byd	byf	byg	byh
can	cao	cap	car	cas	cat	**06**	cav	caw	cae	caz	cab	cac
ceo	cep	cer	ces	cet	cev	**07**	cey	ceu	cez	ceb	cec	ced
cip	cir	cis	cit	civ	ciw	**08**	cie	ciz	cib	cic	cid	cif
cor	cos	cot	cov	cow	coe	**09**	coz	cob	coc	cod	cof	cog
cus	cut	cuv	cuw	cue	cuz	**10**	cub	cuc	cud	cuf	cug	cuh
cyt	cyv	cyw	cye	cyz	cyb	**11**	cyc	cyd	cyf	cyg	cyh	cya
dao	dap	dar	das	dat	dav	**12**	daw	dae	daz	dab	dac	dad
dep	der	des	det	dev	dey	**13**	deu	dez	deb	dec	ded	def
dir	dis	dit	div	diw	die	**14**	diz	dib	dic	did	dif	dig
dos	dot	dov	dow	doe	doz	**15**	dob	doc	dod	dof	dog	doh
dut	duv	duw	due	duz	dub	**16**	duc	dud	duf	dug	duh	dua
dyv	dyw	dye	dyz	dyb	dyc	**17**	dyd	dyf	dyg	dyh	dya	dyk
fap	far	fas	fat	fav	faw	**18**	fae	faz	fab	fac	fad	faf
fer	fes	fet	fev	fey	feu	**19**	fez	feb	fec	fed	fef	feg
fis	fit	fiv	fiw	fie	fiz	**20**	fib	fic	fid	fif	fig	fih
fot	fov	fow	foe	foz	fob	**21**	foc	fod	fof	fog	foh	foa
fuv	fuw	fue	fuz	fub	fuc	**22**	fud	fuf	fug	fuh	fua	fuk
fyw	fye	fyz	fyb	fyc	fyd	**23**	fyf	fyg	fyh	fya	fyk	fyl
gar	gas	gat	gav	gaw	gae	**24**	gaz	gab	gac	gad	gaf	gag
ges	get	gev	gey	geu	gez	**25**	geb	gec	ged	gef	geg	geh
git	giv	giw	gie	giz	gib	**26**	gic	gid	gif	gig	gih	gia
gov	gow	goe	goz	gob	goc	**27**	god	gof	gog	goh	goa	gok
guw	gue	guz	gub	guc	gud	**28**	guf	gug	guh	gua	guk	gul
gye	gyz	gyb	gyc	gyd	gyf	**29**	gyg	gyh	gya	gyk	gyl	gym
has	hat	hav	haw	hae	haz	**30**	hab	hac	had	haf	hag	hau
het	hev	hey	heu	hez	heb	**31**	hec	hed	hef	heg	hei	hea
hiv	hiw	hie	hiz	hib	hic	**32**	hid	hif	hig	hiu	hia	hik
how	hoe	hoz	hob	hoc	hod	**33**	hof	hog	hoi	hoa	hok	hol
hue	huz	hub	huc	hud	huf	**34**	hug	hui	hua	huk	hul	hum
hyz	hyb	hyc	hyd	hyf	hyg	**35**	hyu	hya	hyk	hyl	hym	hyn
kav	kaw	kae	kaz	kab	kac	**36**	kad	kaf	kag	kah	kai	kak
key	keu	kez	keb	kec	ked	**37**	kef	keg	keh	kea	kek	kel
kie	kiz	kib	kic	kid	kif	**38**	kig	kih	kia	kik	kil	kim
koz	kob	koc	kod	kof	kog	**39**	koh	koa	kok	kol	kom	kon
kub	kuc	kud	kuf	kug	kuh	**40**	kua	kuk	kul	kum	kun	kuo
kyc	kyd	kyf	kyg	kyh	kya	**41**	kyk	kyl	kym	kyn	kyo	kyp
law	lae	laz	lab	lac	lad	**42**	laf	lag	lah	lai	lak	lal
leu	lez	leb	lec	led	lef	**43**	leg	leh	lea	lek	lel	lem
liz	lib	lic	lid	lif	lig	**44**	lih	lia	lik	lil	lim	lin
lob	loc	lod	lof	log	loh	**45**	loa	lok	lol	lom	lon	loy
luc	lud	luf	lug	luh	lua	**46**	luk	lul	lum	lun	luo	lup
lyd	lyf	lyg	lyh	lya	lyk	**47**	lyl	lym	lyn	lyo	lyp	lyr
mae	maz	mab	mac	mad	maf	**48**	mag	mah	mai	mak	mal	mam
mez	meb	mec	med	mef	meg	**49**	meh	mea	mek	mel	mem	men

A

First 2 of the 5 letters	ap	el	in	or	ut	ta	re	xo	vu	dy
	80	81	82	83	84	85	86	87	88	89

last 3 of the 5 letters:

1st	2nd	3rd	4th	5th	6th		7th	8th	9th	10th	11th	12th class
mib	mic	mid	mif	mig	mih	**50**	mia	mik	mil	mim	min	mio
moc	mod	mof	mog	moh	moa	**51**	mok	mol	mom	mon	moy	mop
mud	muf	mug	muh	mua	muk	**52**	mul	mum	mun	muo	mup	mur
myf	myg	myh	mya	myk	myl	**53**	mym	myn	myo	myp	myr	mys
naz	nab	nac	nad	naf	nag	**54**	nah	nai	nak	nal	nam	nan
neb	nec	ned	nef	neg	neh	**55**	nea	nek	nel	nem	nen	neo
nic	nid	nif	nig	nih	nia	**56**	nik	nil	nim	nin	nio	nip
nod	nof	nog	noh	noa	nok	**57**	nol	nom	non	noy	nop	nor
nuf	nug	nuh	nua	nuk	nul	**58**	num	nun	nuo	nup	nur	nus
nyg	nyh	nya	nyk	nyl	nym	**59**	nyn	nyo	nyp	nyr	nys	nyt
pac	pad	paf	pag	pah	pai	**60**	pak	pal	pam	pan	pao	pag
ped	pef	peg	peh	pea	pek	**61**	pel	pem	pen	peo	pep	per
pif	pig	pih	pia	pik	pil	**62**	pim	pin	pio	pip	pir	pis
pog	poh	poa	pok	pol	pom	**63**	pon	poy	pop	por	pos	pot
puh	pua	puk	pul	pum	pun	**64**	puo	pup	pur	pus	put	puv
pya	pyk	pyl	pym	pyn	pyo	**65**	pyp	pyr	pys	pyt	pyv	pyw
rad	raf	rag	rah	rai	rak	**66**	ral	ram	ran	rao	rap	rar
ref	reg	reh	rea	rek	rel	**67**	rem	ren	reo	rep	rer	res
rig	rih	ria	rik	ril	rim	**68**	rin	rio	rip	rir	ris	rit
roh	roa	rok	rol	rom	ron	**69**	roy	rop	ror	ros	rot	rov
rua	ruk	rul	rum	run	ruo	**70**	rup	rur	rus	rut	ruv	ruw
ryk	ryl	rym	ryn	ryo	ryp	**71**	ryr	rys	ryt	ryv	ryw	rye
saf	sag	sah	sai	sak	sal	**72**	sam	san	sao	sap	sar	sas
seg	seh	sea	sek	sel	sem	**73**	sen	seo	sep	ser	ses	set
sih	sia	sik	sil	sim	sin	**74**	sio	sip	sir	sis	sit	siv
soa	sok	sol	som	son	soy	**75**	sop	sor	sos	sot	sov	sow
suk	sul	sum	sun	suo	sup	**76**	sur	sus	sut	suv	suw	sue
syl	sym	syn	syo	syp	syr	**77**	sys	syt	syv	syw	sye	syz
tag	tah	tai	tak	tal	tam	**78**	tan	tao	tap	tar	tas	tat
teh	tea	tek	tel	tem	ten	**79**	teo	tep	ter	tes	tet	tev
tia	tik	til	tim	tin	tio	**80**	tip	tir	tis	tit	tiv	tiw
tok	tol	tom	ton	toy	top	**81**	tor	tos	tot	tov	tow	toe
tul	tum	tun	tuo	tup	tur	**82**	tus	tut	tuv	tuw	tue	tuz
tym	tyn	tyo	typ	tyr	tys	**83**	tyt	tyv	tyw	tye	tyz	tyb
vah	vai	vak	val	vam	van	**84**	vao	vap	var	vas	vat	vav
vea	vek	vel	vem	ven	veo	**85**	vep	ver	ves	vet	vev	vey
vik	vil	vim	vin	vio	vip	**86**	vir	vis	vit	viv	viw	vie
vol	vom	von	voy	vop	vor	**87**	vos	vot	vov	vow	voe	voz
vum	vun	vuo	vup	vur	vus	**88**	vut	vuv	vuw	vue	vuz	vub
vyn	vyo	vyp	vyr	vys	vyt	**89**	vyv	vyw	vye	vyz	vyb	vyc
wai	wak	wal	wam	wan	wao	**90**	wap	war	was	wat	wav	waw
wek	wel	wem	wen	weo	wep	**91**	wer	wes	wet	wev	wey	weu
wil	wim	win	wio	wip	wir	**92**	wis	wit	wiv	woe	wie	wiz
wom	won	woy	wop	wor	wos	**93**	wot	wov	wue	wuz	woz	wob
zal	zam	zan	zao	zap	zar	**94**	zas	zat	zav	zaw	zae	zaz
zem	zen	zeo	zep	zer	zes	**95**	zet	zev	zey	zeu	zez	zeb
zin	zio	zip	zir	zis	zit	**96**	ziv	ziw	zie	ziz	zib	zic
zoy	zop	zor	zos	zot	zov	**97**	zow	zoe	zoz	zob	zoc	zod
zup	zur	zus	zut	zuv	zuw	**98**	zue	zuz	zub	zuc	zud	zuf
zyr	zys	zyt	zyv	zyw	zye	**99**	zyz	zyb	zyc	zyd	zyf	zyg

A
80

A

First 2 of the 5 letters: an/90, ej/91, ir/92, uv/93, wa/94, ze/95, fi/96, po/97, gu/98, my/99

	an	ej	ir	uv	wa	ze	fi	po	gu	my
First 2 of the 5 letters:	90	91	92	93	94	95	96	97	98	99

last 3 of the 5 letters:

	1st	2nd	3rd	4th	5th	6th		7th	8th	9th	10th	11th	12th class
A	ban	bao	bap	bar	bas	bat	**00**	bav	baw	bae	baz	bab	bac
	beo	bep	ber	bes	bet	bev	**01**	bey	beu	bez	beb	bec	bed
90	bip	bir	bis	bit	biv	biw	**02**	bie	biz	bib	bic	bid	bif
	bor	bos	bot	bov	bow	boe	**03**	boz	bob	boc	bod	bof	bog
	bus	but	buv	buw	bue	buz	**04**	bub	buc	bud	buf	bug	buh
	byt	byv	byw	bye	byz	byb	**05**	byc	byd	byf	byg	byh	bya
	cao	cap	car	cas	cat	cav	**06**	caw	cae	caz	cab	cac	cad
	cep	cer	ces	cet	cev	cey	**07**	ceu	cez	ceb	cec	ced	cef
	cir	cis	cit	civ	ciw	cie	**08**	ciz	cib	cic	cid	cif	cig
	cos	cot	cov	cow	coe	coz	**09**	cob	coc	cod	cof	cog	coh
	cut	cuv	cuw	cue	cuz	cub	**10**	cuc	cud	cuf	cug	cuh	cua
	cyv	cyw	cye	cyz	cyb	cyc	**11**	cyd	cyf	cyg	cyh	cya	cyk
	dap	dar	das	dat	dav	daw	**12**	dae	daz	dab	dac	dad	daf
	der	des	det	dev	dey	deu	**13**	dez	deb	dec	ded	def	deg
	dis	dit	div	diw	die	diz	**14**	dib	dic	did	dif	dig	dih
	dot	dov	dow	doe	doz	dob	**15**	doc	dod	dof	dog	doh	doa
	duv	duw	due	duz	dub	duc	**16**	dud	duf	dug	duh	dua	duk
	dyw	dye	dyz	dyb	dyc	dyd	**17**	dyf	dyg	dyh	dya	dyk	dyl
	far	fas	fat	fav	faw	fae	**18**	faz	fab	fac	fad	faf	fag
	fes	fet	fev	fey	feu	fez	**19**	feb	fec	fed	fef	feg	feh
	fit	fiv	fiw	fie	fiz	fib	**20**	fic	fid	fif	fig	fih	fia
	fov	fow	foe	foz	fob	foc	**21**	fod	fof	fog	foh	foa	fok
	fuw	fue	fuz	fub	fuc	fud	**22**	fuf	fug	fuh	fua	fuk	ful
	fye	fyz	fyb	fyc	fyd	fyf	**23**	fyg	fyh	fya	fyk	fyl	fym
	gas	gat	gav	gaw	gae	gaz	**24**	gab	gac	gad	gaf	gag	gah
	get	gev	gey	geu	gez	geb	**25**	gec	ged	gef	geg	geh	gea
	giv	giw	gie	giz	gib	gic	**26**	gid	gif	gig	gih	gia	gik
	gow	goe	goz	gob	goc	god	**27**	gof	gog	goh	goa	gok	gol
	gue	guz	gub	guc	gud	guf	**28**	gug	guh	gua	guk	gul	gum
	gyz	gyb	gyc	gyd	gyf	gyg	**29**	gyh	gya	gyk	gyl	gym	gyn
	hat	hav	haw	hae	haz	hab	**30**	hac	had	haf	hag	hau	hai
	hev	hey	heu	hez	heb	hec	**31**	hed	hef	heg	hei	hea	hek
	hiw	hie	hiz	hib	hic	hid	**32**	hif	hig	hiu	hia	hik	hil
	hoe	hoz	hob	hoc	hod	hof	**33**	hog	hoi	hoa	hok	hol	hom
	huz	hub	huc	hud	huf	hug	**34**	hui	hua	huk	hul	hum	hun
	hyb	hyc	hyd	hyf	hyg	hyu	**35**	hya	hyk	hyl	hym	hyn	hyo
	kaw	kae	kaz	kab	kac	kad	**36**	kaf	kag	kah	kai	kak	kal
	keu	kez	keb	kec	ked	kef	**37**	keg	keh	kea	kek	kel	kem
	kiz	kib	kic	kid	kif	kig	**38**	kih	kia	kik	kil	kim	kin
	kob	koc	kod	kof	kog	koh	**39**	koa	kok	kol	kom	kon	koy
	kuc	kud	kuf	kug	kuh	kua	**40**	kuk	kul	kum	kun	kuo	kup
	kyd	kyf	kyg	kyh	kya	kyk	**41**	kyl	kym	kyn	kyo	kyp	kyr
	lae	laz	lab	lac	lad	laf	**42**	lag	lah	lai	lak	lal	lam
	lez	leb	lec	led	lef	leg	**43**	leh	lea	lek	lel	lem	len
	lib	lic	lid	lif	lig	lih	**44**	lia	lik	lil	lim	lin	lio
	loc	lod	lof	log	loh	loa	**45**	lok	lol	lom	lon	loy	lop
	lud	luf	lug	luh	lua	luk	**46**	lul	lum	lun	luo	lup	lur
	lyf	lyg	lyh	lya	lyk	lyl	**47**	lym	lyn	lyo	lyp	lyr	lys
	maz	mab	mac	mad	maf	mag	**48**	mah	mai	mak	mal	mam	man
	meb	mec	med	mef	meg	meh	**49**	mea	mek	mel	mem	men	meo

A

First 2 of the 5 letters:	an	ej	ir	uv	wa	ze	fi	po	gu	my
	90	91	92	93	94	95	96	97	98	99

last 3 of the 5 letters:

1st	2nd	3rd	4th	5th	6th		7th	8th	9th	10th	11th	12th class
mic	**mid**	**mif**	**mig**	**mih**	**mia**	**50**	**mik**	**mil**	**mim**	**min**	**mio**	**mip**
mod	mof	mog	moh	moa	mok	**51**	mol	mom	mon	moy	mop	mor
muf	mug	muh	mua	muk	mul	**52**	mum	mun	muo	mup	mur	mus
myg	myh	mya	myk	myl	mym	**53**	myn	myo	myp	myr	mys	myt
nab	**nac**	**nad**	**naf**	**nag**	**nah**	**54**	**nai**	**nak**	**nal**	**nam**	**nan**	**nao**
nec	ned	nef	neg	neh	nea	**55**	nek	nel	nem	nen	neo	nep
nid	nif	nig	nih	nia	nik	**56**	nil	nim	nin	nio	nip	nir
nof	nog	noh	noa	nok	nol	**57**	nom	non	noy	nop	nor	nos
nug	nuh	nua	nuk	nul	num	**58**	nun	nuo	nup	nur	nus	nut
nyh	nya	nyk	nyl	nym	nyn	**59**	nyo	nyp	nyr	nys	nyt	nyv
pad	**paf**	**pag**	**pah**	**pai**	**pak**	**60**	**pal**	**pam**	**pan**	**pao**	**pap**	**par**
pef	peg	peh	pea	pek	pel	**61**	pem	pen	peo	pep	per	pes
pig	pih	pia	pik	pil	pim	**62**	pin	pio	pip	pir	pis	pit
poh	poa	pok	pol	pom	pon	**63**	poy	pop	por	pos	pot	pov
pua	puk	pul	pum	pun	puo	**64**	pup	pur	pus	put	puv	puw
pyk	pyl	pym	pyn	pyo	pyp	**65**	pyr	pys	pyt	pyv	pyw	pye
raf	**rag**	**rah**	**rai**	**rak**	**ral**	**66**	**ram**	**ran**	**rao**	**rap**	**rar**	**ras**
reg	reh	rea	rek	rel	rem	**67**	ren	reo	rep	rer	res	ret
rih	ria	rik	ril	rim	rin	**68**	rio	rip	rir	ris	rit	riv
roa	rok	rol	rom	ron	roy	**69**	rop	ror	ros	rot	rov	row
ruk	rul	rum	run	ruo	rup	**70**	rur	rus	rut	ruv	ruw	rue
ryl	rym	ryn	ryo	ryp	ryr	**71**	rys	ryt	ryv	ryw	rye	ryz
sag	**sah**	**sai**	**sak**	**sal**	**sam**	**72**	**san**	**sao**	**sap**	**sar**	**sas**	**sat**
seh	sea	sek	sel	sem	sen	**73**	seo	sep	ser	ses	set	sev
sia	sik	sil	sim	sin	sio	**74**	sip	sir	sis	sit	siv	siw
sok	sol	som	son	soy	sop	**75**	sor	sos	sot	sov	sow	soe
sul	sum	sun	suo	sup	sur	**76**	sus	sut	suv	suw	sue	suz
sym	syn	syo	syp	syr	sys	**77**	syt	syv	syw	sye	syz	syb
tah	**tai**	**tak**	**tal**	**tam**	**tan**	**78**	**tao**	**tap**	**tar**	**tas**	**tat**	**tav**
tea	tek	tel	tem	ten	teo	**79**	tep	ter	tes	tet	tev	tey
tik	til	tim	tin	tio	tip	**80**	tir	tis	tit	tiv	tiw	tie
tol	tom	ton	toy	top	tor	**81**	tos	tot	tov	tow	toe	toz
tum	tun	tuo	tup	tur	tus	**82**	tut	tuv	tuw	tue	tuz	tub
tyn	tyo	typ	tyr	tys	tyt	**83**	tyv	tyw	tye	tyz	tyb	tyc
vai	**vak**	**val**	**vam**	**van**	**vao**	**84**	**vap**	**var**	**vas**	**vat**	**vav**	**vaw**
vek	vel	vem	ven	veo	vep	**85**	ver	ves	vet	vev	vey	veu
vil	vim	vin	vio	vip	vir	**86**	vis	vit	viv	viw	vie	viz
vom	von	voy	vop	vor	vos	**87**	vot	vov	vow	voe	voz	vob
vun	vuo	vup	vur	vus	vut	**88**	vuv	vuw	vue	vuz	vub	vuc
vyo	vyp	vyr	vys	vyt	vyv	**89**	vyw	vye	vyz	vyb	vyc	vyd
wak	**wal**	**wam**	**wan**	**wao**	**wap**	**90**	**war**	**was**	**wat**	**wav**	**waw**	**wae**
wel	wem	wen	weo	wep	wer	**91**	wes	wet	wev	wey	weu	wez
wim	win	wio	wip	wir	wis	**92**	wit	wiv	woe	wie	wiz	wib
won	woy	wop	wor	wos	wot	**93**	wov	wue	wuz	woz	wob	woc
zam	**zan**	**zao**	**zap**	**zar**	**zas**	**94**	**zat**	**zav**	**zaw**	**zae**	**zag**	**zab**
zen	zeo	zep	zer	zes	zet	**95**	zev	zey	zeu	zez	zeb	zec
zio	zip	zir	zis	zit	ziv	**96**	ziw	zie	ziz	zib	zic	zid
zop	zor	zos	zot	zov	zow	**97**	zoe	zoz	zob	zoc	zod	zof
zur	zus	zut	zuv	zuw	zue	**98**	zuz	zub	zuc	zud	zuf	zug
zys	zyt	zyv	zyw	zye	zyz	**99**	zyb	zyc	zyd	zyf	zyg	zyh

A 90

B

First 2 of the 5 letters:	az	ed	oj	us	va	ce	xi	do	tu	fy
	00	01	02	03	04	05	06	07	08	09

last 3 of the 5 letters:

B
00

1st	2nd	3rd	4th	5th	6th		7th	8th	9th	10th	11th	12th class
bao	bap	bar	bas	bat	bav	00	baw	bae	baz	bab	bac	bad
bep	ber	bes	bet	bev	bey	01	beu	bez	beb	bec	bed	bef
bir	bis	bit	biv	biw	bie	02	biz	bib	bic	bid	bif	big
bos	bot	bov	bow	boe	boz	03	bob	boc	bod	bof	bog	boh
but	buv	buw	bue	buz	bub	04	buc	bud	buf	bug	buh	bua
byv	byw	bye	byz	byb	byc	05	byd	byf	byg	byh	bya	byk
cap	car	cas	cat	cav	caw	06	cae	caz	cab	cac	cad	caf
cer	ces	cet	cev	cey	ceu	07	cez	ceb	cec	ced	cef	ceg
cis	cit	civ	ciw	cie	ciz	08	cib	cic	cid	cif	cig	cih
cot	cov	cow	coe	coz	cob	09	coc	cod	cof	cog	coh	coa
cuv	cuw	cue	cuz	cub	cuc	10	cud	cuf	cug	cuh	cua	cuk
cyw	cye	cyz	cyb	cyc	cyd	11	cyf	cyg	cyh	cya	cyk	cyl
dar	das	dat	dav	daw	dae	12	daz	dab	dac	dad	daf	dag
des	det	dev	dey	deu	dez	13	deb	dec	ded	def	deg	deh
dit	div	diw	die	diz	dib	14	dic	did	dif	dig	dih	dia
dov	dow	doe	doz	dob	doc	15	dod	dof	dog	doh	doa	dok
duw	due	duz	dub	duc	dud	16	duf	dug	duh	dua	duk	dul
dye	dyz	dyb	dyc	dyd	dyf	17	dyg	dyh	dya	dyk	dyl	dym
fas	fat	fav	faw	fae	faz	18	fab	fac	fad	faf	fag	fah
fet	fev	fey	feu	fez	feb	19	fec	fed	fef	feg	feh	fea
fiv	fiw	fie	fiz	fib	fic	20	fid	fif	fig	fih	fia	fik
fow	foe	foz	fob	foc	fod	21	fof	fog	foh	foa	fok	fol
fue	fuz	fub	fuc	fud	fuf	22	fug	fuh	fua	fuk	ful	fum
fyz	fyb	fyc	fyd	fyf	fyg	23	fyh	fya	fyk	fyl	fym	fyn
gat	gav	gaw	gae	gaz	gab	24	gac	gad	gaf	gag	gah	gai
gev	gey	geu	gez	geb	gec	25	ged	gef	geg	geh	gea	gek
giw	gie	giz	gib	gic	gid	26	gif	gig	gih	gia	gik	gil
goe	goz	gob	goc	god	gof	27	gog	goh	goa	gok	gol	gom
guz	gub	guc	gud	guf	gug	28	guh	gua	guk	gul	gum	gun
gyb	gyc	gyd	gyf	gyg	gyh	29	gya	gyk	gyl	gym	gyn	gyo
hav	haw	hae	haz	hab	hac	30	had	haf	hag	hau	hai	hak
hey	heu	hez	heb	hec	hed	31	hef	heg	hei	hea	hek	hel
hie	hiz	hib	hic	hid	hif	32	hig	hiu	hia	hik	hil	him
hoz	hob	hoc	hod	hof	hog	33	hoi	hoa	hok	hol	hom	hon
hub	huc	hud	huf	hug	hui	34	hua	huk	hul	hum	hun	huo
hyc	hyd	hyf	hyg	hyu	hya	35	hyk	hyl	hym	hyn	hyo	hyp
kae	kaz	kab	kac	kad	kaf	36	kag	kah	kai	kak	kal	kam
kez	keb	kec	ked	kef	keg	37	keh	kea	kek	kel	kem	ken
kib	kic	kid	kif	kig	kih	38	kia	kik	kil	kim	kin	kio
koc	kod	kof	kog	koh	koa	39	kok	kol	kom	kon	koy	kop
kud	kuf	kug	kuh	kua	kuk	40	kul	kum	kun	kuo	kup	kur
kyf	kyg	kyh	kya	kyk	kyl	41	kym	kyn	kyo	kyp	kyr	kys
laz	lab	lac	lad	laf	lag	42	lah	lai	lak	lal	lam	lan
leb	lec	led	lef	leg	leh	43	lea	lek	lel	lem	len	leo
lic	lid	lif	lig	lih	lia	44	lik	lil	lim	lin	lio	lip
lod	lof	log	loh	loa	lok	45	lol	lom	lon	loy	lop	lor
luf	lug	luh	lua	luk	lul	46	lum	lun	luo	lup	lur	lus
lyg	lyh	lya	lyk	lyl	lym	47	lyn	lyo	lyp	lyr	lys	lyt
mab	mac	mad	maf	mag	mah	48	mai	mak	mal	mam	man	mao
mec	med	mef	meg	meh	mea	49	mek	mel	mem	men	meo	mep

B

First 2 of the 5 letters: { **az** 00 | **ed** 01 | **oj** 02 | **us** 03 | **va** 04 | **ce** 05 | **xi** 06 | **do** 07 | **tu** 08 | **fy** 09

last 3 of the 5 letters:

1st	2nd	3rd	4th	5th	6th		7th	8th	9th	10th	11th	12th class
mid	mif	mig	mih	mia	mik	50	mil	mim	min	mio	mip	mir
mof	mog	moh	moa	mok	mol	51	mom	mon	moy	mop	mor	mos
mug	muh	mua	muk	mul	mum	52	mun	muo	mup	mur	mus	mut
myh	mya	myk	myl	mym	myn	53	myo	myp	myr	mys	myt	myv
nac	nad	naf	nag	nah	nai	54	nak	nal	nam	nan	nao	nap
ned	nef	neg	neh	nea	nek	55	nel	nem	nen	neo	nep	ner
nif	nig	nih	nia	nik	nil	56	nim	nin	nio	nip	nir	nis
nog	noh	noa	nok	nol	nom	57	non	noy	nop	nor	nos	not
nuh	nua	nuk	nul	num	nun	58	nuo	nup	nur	nus	nut	nuv
nya	nyk	nyl	nym	nyn	nyo	59	nyp	nyr	nys	nyt	nyv	nyw
paf	pag	pah	pai	pak	pal	60	pam	pan	pao	pap	par	pas
peg	peh	pea	pek	pel	pem	61	pen	peo	pep	per	pes	pet
pih	pia	pik	pil	pim	pin	62	pio	pip	pir	pis	pit	piv
poa	pok	pol	pom	pon	poy	63	pop	por	pos	pot	pov	pow
puk	pul	pum	pun	puo	pup	64	pur	pus	put	puv	puw	pue
pyl	pym	pyn	pyo	pyp	pyr	65	pys	pyt	pyv	pyw	pye	pyz
rag	rah	rai	rak	ral	ram	66	ran	rao	rap	rar	ras	rat
reh	rea	rek	rel	rem	ren	67	reo	rep	rer	res	ret	rev
ria	rik	ril	rim	rin	rio	68	rip	rir	ris	rit	riv	riw
rok	rol	rom	ron	roy	rop	69	ror	ros	rot	rov	row	roe
rul	rum	run	ruo	rup	rur	70	rus	rut	ruv	ruw	rue	ruz
rym	ryn	ryo	ryp	ryr	rys	71	ryt	ryv	ryw	rye	ryz	ryb
sah	sai	sak	sal	sam	san	72	sao	sap	sar	sas	sat	sav
sea	sek	sel	sem	sen	seo	73	sep	ser	ses	set	sev	sey
sik	sil	sim	sin	sio	sip	74	sir	sis	sit	siv	siw	sie
sol	som	son	soy	sop	sor	75	sos	sot	sov	sow	soe	soz
sum	sun	suo	sup	sur	sus	76	sut	suv	suw	sue	suz	sub
syn	syo	syp	syr	sys	syt	77	syv	syw	sye	syz	syb	syc
tai	tak	tal	tam	tan	tao	78	tap	tar	tas	tat	tav	taw
tek	tel	tem	ten	teo	tep	79	ter	tes	tet	tev	tey	teu
til	tim	tin	tio	tip	tir	80	tis	tit	tiv	tiw	tie	tiz
tom	ton	toy	top	tor	tos	81	tot	tov	tow	toe	toz	tob
tun	tuo	tup	tur	tus	tut	82	tuv	tuw	tue	tuz	tub	tuc
tyo	typ	tyr	tys	tyt	tyv	83	tyw	tye	tyz	tyb	tyc	tyd
vak	val	vam	van	vao	vap	84	var	vas	vat	vav	vaw	vae
vel	vem	ven	veo	vep	ver	85	ves	vet	vev	vey	veu	vez
vim	vin	vio	vip	vir	vis	86	vit	viv	viw	vie	viz	vib
von	voy	vop	vor	vos	vot	87	vov	vow	voe	voz	vob	voc
vuo	vup	vur	vus	vut	vuv	88	vuw	vue	vuz	vub	vuc	vud
vyp	vyr	vys	vyt	vyv	vyw	89	vye	vyz	vyb	vyc	vyd	vyf
wal	wam	wan	wao	wap	war	90	was	wat	wav	waw	wae	waz
wem	wen	weo	wep	wer	wes	91	wet	wev	wey	weu	wez	web
win	wio	wip	wir	wis	wit	92	wiv	woe	wie	wiz	wib	wic
woy	wop	wor	wos	wot	wov	93	wue	wuz	woz	wob	woc	wod
zan	zao	zap	zar	zas	zat	94	zav	zaw	zae	zaz	zab	zac
zeo	zep	zer	zes	zet	zev	95	zey	zeu	zez	zeb	zec	zed
zip	zir	zis	zit	ziv	ziw	96	zie	ziz	zib	zic	zid	zif
zor	zos	zot	zov	zow	zoe	97	zoz	zob	zoc	zod	zof	zog
zus	zut	zuv	zuw	zue	zuz	98	zub	zuc	zud	zuf	zug	zuh
zyt	zyv	zyw	zye	zyz	zyb	99	zyc	zyd	zyf	zyg	zyh	zya

B

First 2 of the 5 letters	am	ez	iv	ob	la	xe	ti	fo	wu	ky
	10	11	12	13	14	15	16	17	18	19

last 3 of the 5 letters:

B
10

1st	2nd	3rd	4th	5th	6th		7th	8th	9th	10th	11th	12th class
bap	bar	bas	bat	bav	baw	00	bae	baz	bab	bac	bad	baf
ber	bes	bet	bev	bey	beu	01	bez	beb	bec	bed	bef	beg
bis	bit	biv	biw	bie	biz	02	bib	bic	bid	bif	big	bih
bot	bov	bow	boe	boz	bob	03	boc	bod	bof	bog	boh	boa
buv	buw	bue	buz	bub	buc	04	bud	buf	bug	buh	bua	buk
byw	bye	byz	byb	byc	byd	05	byf	byg	byh	bya	byk	byl
car	cas	cat	cav	caw	cae	06	caz	cab	cac	cad	caf	cag
ces	cet	cev	cey	ceu	cez	07	ceb	cec	ced	cef	ceg	ceh
cit	civ	ciw	cie	ciz	cib	08	cic	cid	cif	cig	cih	cia
cov	cow	coe	coz	cob	coc	09	cod	cof	cog	coh	coa	cok
cuw	cue	cuz	cub	cuc	cud	10	cuf	cug	cuh	cua	cuk	cul
cye	cyz	cyb	cyc	cyd	cyf	11	cyg	cyh	cya	cyk	cyl	cym
das	dat	dav	daw	dae	daz	12	dab	dac	dad	daf	dag	dah
det	dev	dey	deu	dez	deb	13	dec	ded	def	deg	deh	dea
div	diw	die	diz	dib	dic	14	did	dif	dig	dih	dia	dik
dow	doe	doz	dob	doc	dod	15	dof	dog	doh	doa	dok	dol
due	duz	dub	duc	dud	duf	16	dug	duh	dua	duk	dul	dum
dyz	dyb	dyc	dyd	dyf	dyg	17	dyh	dya	dyk	dyl	dym	dyn
fat	fav	faw	fae	faz	fab	18	fac	fad	faf	fag	fah	fai
fev	fey	feu	fez	feb	fec	19	fed	fef	feg	feh	fea	fek
fiw	fie	fiz	fib	fic	fid	20	fif	fig	fih	fia	fik	fil
foe	foz	fob	foc	fod	fof	21	fog	foh	foa	fok	fol	fom
fuz	fub	fuc	fud	fuf	fug	22	fuh	fua	fuk	ful	fum	fun
fyb	fyc	fyd	fyf	fyg	fyh	23	fya	fyk	fyl	fym	fyn	fyo
gav	gaw	gae	gaz	gab	gac	24	gad	gaf	gag	gah	gai	gak
gey	geu	gez	geb	gec	ged	25	gef	geg	geh	gea	gek	gel
gie	giz	gib	gic	gid	gif	26	gig	gih	gia	gik	gil	gim
goz	gob	goc	god	gof	gog	27	goh	goa	gok	gol	gom	gon
gub	guc	gud	guf	gug	guh	28	gua	guk	gul	gum	gun	guo
gyc	gyd	gyf	gyg	gyh	gya	29	gyk	gyl	gym	gyn	gyo	gyp
haw	hae	haz	hab	hac	had	30	haf	hag	hau	hai	hak	hal
heu	hez	heb	hec	hed	hef	31	heg	hei	hea	hek	hel	hem
hiz	hib	hic	hid	hif	hig	32	hiu	hia	hik	hil	him	hin
hob	hoc	hod	hof	hog	hoi	33	hoa	hok	hol	hom	hon	hoy
huc	hud	huf	hug	hui	hua	34	huk	hul	hum	hun	huo	hup
hyd	hyf	hyg	hyu	hya	hyk	35	hyl	hym	hyn	hyo	hyp	hyr
kaz	kab	kac	kad	kaf	kag	36	kah	kai	kak	kal	kam	kan
keb	kec	ked	kef	keg	keh	37	kea	kek	kel	kem	ken	keo
kic	kid	kif	kig	kih	kia	38	kik	kil	kim	kin	kio	kip
kod	kof	kog	koh	koa	kok	39	kol	kom	kon	koy	kop	kor
kuf	kug	kuh	kua	kuk	kul	40	kum	kun	kuo	kup	kur	kus
kyg	kyh	kya	kyk	kyl	kym	41	kyn	kyo	kyp	kyr	kys	kyt
lab	lac	lad	laf	lag	lah	42	lai	lak	lal	lam	lan	lao
lec	led	lef	leg	leh	lea	43	lek	lel	lem	len	leo	lep
lid	lif	lig	lih	lia	lik	44	lil	lim	lin	lio	lip	lir
lof	log	loh	loa	lok	lol	45	lom	lon	loy	lop	lor	los
lug	luh	lua	luk	lul	lum	46	lun	luo	lup	lur	lus	lut
lyh	lya	lyk	lyl	lym	lyn	47	lyo	lyp	lyr	lys	lyt	lyv
mac	mad	maf	mag	mah	mai	48	mak	mal	mam	man	mao	map
med	mef	meg	meh	mea	mek	49	mel	mem	men	meo	mep	mer

B

First 2 of the 5 letters	am	ez	iv	ob	la	xe	ti	fo	wu	ky
	10	11	12	13	14	15	16	17	18	19

last 3 of the 5 letters:

1st	2nd	3rd	4th	5th	6th		7th	8th	9th	10th	11th	12th class
mif	mig	mih	mia	mik	mil	**50**	mim	min	mio	mip	mir	mis
mog	moh	moa	mok	mol	mom	**51**	mon	moy	mop	mor	mos	mot
muh	mua	muk	mul	mum	mun	**52**	muo	mup	mur	mus	mut	muv
mya	myk	myl	mym	myn	myo	**53**	myp	myr	mys	myt	myv	myw
nad	naf	nag	nah	nai	nak	**54**	nal	nam	nan	nao	nap	nar
nef	neg	neh	nea	nek	nel	**55**	nem	nen	neo	nep	ner	nes
nig	nih	nia	nik	nil	nim	**56**	nin	nio	nip	nir	nis	nit
noh	noa	nok	nol	nom	non	**57**	noy	nop	nor	nos	not	nov
nua	nuk	nul	num	nun	nuo	**58**	nup	nur	nus	nut	nuv	nuw
nyk	nyl	nym	nyn	nyo	nyp	**59**	nyr	nys	nyt	nyv	nyw	nye
pag	pah	pai	pak	pal	pam	**60**	pan	pao	pap	par	pas	pat
peh	pea	pek	pel	pem	pen	**61**	peo	pep	per	pes	pet	pev
pia	pik	pil	pim	pin	pio	**62**	pip	pir	pis	pit	piv	piw
pok	pol	pom	pon	poy	pop	**63**	por	pos	pot	pov	pow	poe
pul	pum	pun	puo	pup	pur	**64**	pus	put	puv	puw	pue	puz
pym	pyn	pyo	pyp	pyr	pys	**65**	pyt	pyv	pyw	pye	pyz	pyb
rah	rai	rak	ral	ram	ran	**66**	rao	rap	rar	ras	rat	rav
rea	rek	rel	rem	ren	reo	**67**	rep	rer	res	ret	rev	rey
rik	ril	rim	rin	rio	rip	**68**	rir	ris	rit	riv	riw	rie
rol	rom	ron	roy	rop	ror	**69**	ros	rot	rov	row	roe	roz
rum	run	ruo	rup	rur	rus	**70**	rut	ruv	ruw	rue	ruz	rub
ryn	ryo	ryp	ryr	rys	ryt	**71**	ryv	ryw	rye	ryz	ryb	ryc
sai	sak	sal	sam	san	sao	**72**	sap	sar	sas	sat	sav	saw
sek	sel	sem	sen	seo	sep	**73**	ser	ses	set	sev	sey	seu
sil	sim	sin	sio	sip	sir	**74**	sis	sit	siv	siw	sie	siz
som	son	soy	sop	sor	sos	**75**	sot	sov	sow	soe	soz	sob
sun	suo	sup	sur	sus	sut	**76**	suv	suw	sue	suz	sub	suc
syo	syp	syr	sys	syt	syv	**77**	syw	sye	syz	syb	syc	syd
tak	tal	tam	tan	tao	tap	**78**	tar	tas	tat	tav	taw	tae
tel	tem	ten	teo	tep	ter	**79**	tes	tet	tev	tey	teu	tez
tim	tin	tio	tip	tir	tis	**80**	tit	tiv	tiw	tie	tiz	tib
ton	toy	top	tor	tos	tot	**81**	tov	tow	toe	toz	tob	toc
tuo	tup	tur	tus	tut	tuv	**82**	tuw	tue	tuz	tub	tuc	tud
typ	tyr	tys	tyt	tyv	tyw	**83**	tye	tyz	tyb	tyc	tyd	tyf
val	vam	van	vao	vap	var	**84**	vas	vat	vav	vaw	vae	vaz
vem	ven	veo	vep	ver	ves	**85**	vet	vev	vey	veu	vez	veb
vin	vio	vip	vir	vis	vit	**86**	viv	viw	vie	viz	vib	vic
voy	vop	vor	vos	vot	vov	**87**	vow	voe	voz	vob	voc	vod
vup	vur	vus	vut	vuv	vuw	**88**	vue	vuz	vub	vuc	vud	vuf
vyr	vys	vyt	vyv	vyw	vye	**89**	vyz	vyb	vyc	vyd	vyf	vyg
wam	wan	wao	wap	war	was	**90**	wat	wav	waw	wae	waz	wab
wen	weo	wep	wer	wes	wet	**91**	wev	wey	weu	wez	web	wec
wio	wip	wir	wis	wit	wiv	**92**	woe	wie	wiz	wib	wic	wid
wop	wor	wos	wot	wov	wue	**93**	wuz	woz	wob	woc	wod	wof
zao	zap	zar	zas	zat	zav	**94**	zaw	zae	zaz	zab	zac	zad
zep	zer	zes	zet	zev	zey	**95**	zeu	zez	zeb	zec	zed	zef
zir	zis	zit	ziv	ziw	zie	**96**	ziz	zib	zic	zid	zif	zig
zos	zot	zov	zow	zoe	zoz	**97**	zob	zoc	zod	zof	zog	zoh
zut	zuv	zuw	zue	zuz	zub	**98**	zuc	zud	zuf	zug	zuh	zua
zyv	zyw	zye	zyz	zyb	zyc	**99**	zyd	zyf	zyg	zyh	zya	zyk

B

First 2 of the 5 letters:	ax	ec	ig	ok	uw	ca	di	go	lu	ny
	20	21	22	23	24	25	26	27	28	29

last 3 of the 5 letters:

	1st	2nd	3rd	4th	5th	6th		7th	8th	9th	10th	11th	12th class
B	bar	bas	bat	bav	baw	bae	00	baz	bab	bac	bad	baf	bag
	bes	bet	bev	bey	beu	bez	01	beb	bec	bed	bef	beg	beh
20	bit	biv	biw	bie	biz	bib	02	bic	bid	bif	big	bih	bia
	bov	bow	boe	boz	bob	boc	03	bod	bof	bog	boh	boa	bok
	buw	bue	buz	bub	buc	bud	04	buf	bug	buh	bua	buk	bul
	bye	byz	byb	byc	byd	byf	05	byg	byh	bya	byk	byl	bym
	cas	cat	cav	caw	cae	caz	06	cab	cac	cad	caf	cag	cah
	cet	cev	cey	ceu	cez	ceb	07	cec	ced	cef	ceg	ceh	cea
	civ	ciw	cie	ciz	cib	cic	08	cid	cif	cig	cih	cia	cik
	cow	coe	coz	cob	coc	cod	09	cof	cog	coh	coa	cok	col
	cue	cuz	cub	cuc	cud	cuf	10	cug	cuh	cua	cuk	cul	cum
	cyz	cyb	cyc	cyd	cyf	cyg	11	cyh	cya	cyk	cyl	cym	cyn
	dat	dav	daw	dae	daz	dab	12	dac	dad	daf	dag	dah	dai
	dev	dey	deu	dez	deb	dec	13	ded	def	deg	deh	dea	dek
	diw	die	diz	dib	dic	did	14	dif	dig	dih	dia	dik	dil
	doe	doz	dob	doc	dod	dof	15	dog	doh	doa	dok	dol	dom
	duz	dub	duc	dud	duf	dug	16	duh	dua	duk	dul	dum	dun
	dyb	dyc	dyd	dyf	dyg	dyh	17	dya	dyk	dyl	dym	dyn	dyo
	fav	faw	fae	faz	fab	fac	18	fad	faf	fag	fah	fai	fak
	fey	feu	fez	feb	fec	fed	19	fef	feg	feh	fea	fek	fel
	fie	fiz	fib	fic	fid	fif	20	fig	fih	fia	fik	fil	fim
	foz	fob	foc	fod	fof	fog	21	foh	foa	fok	fol	fom	fon
	fub	fuc	fud	fuf	fug	fuh	22	fua	fuk	ful	fum	fun	fuo
	fyc	fyd	fyf	fyg	fyh	fya	23	fyk	fyl	fym	fyn	fyo	fyp
	gaw	gae	gaz	gab	gac	gad	24	gaf	gag	gah	gai	gak	gal
	geu	gez	geb	gec	ged	gef	25	geg	geh	gea	gek	gel	gem
	giz	gib	gic	gid	gif	gig	26	gih	gia	gik	gil	gim	gin
	gob	goc	god	gof	gog	goh	27	goa	gok	gol	gom	gon	goy
	guc	gud	guf	gug	guh	gua	28	guk	gul	gum	gun	guo	gup
	gyd	gyf	gyg	gyh	gya	gyk	29	gyl	gym	gyn	gyo	gyp	gyr
	hae	haz	hab	hac	had	haf	30	hag	hau	hai	hak	hal	ham
	hez	heb	hec	hed	hef	heg	31	hei	hea	hek	hel	hem	hen
	hib	hic	hid	hif	hig	hiu	32	hia	hik	hil	him	hin	hio
	hoc	hod	hof	hog	hoi	hoa	33	hok	hol	hom	hon	hoy	hop
	hud	huf	hug	hui	hua	huk	34	hul	hum	hun	huo	hup	hur
	hyf	hyg	hyu	hya	hyk	hyl	35	hym	hyn	hyo	hyp	hyr	hys
	kab	kac	kad	kaf	kag	kah	36	kai	kak	kal	kam	kan	kao
	kec	ked	kef	keg	keh	kea	37	kek	kel	kem	ken	keo	kep
	kid	kif	kig	kih	kia	kik	38	kil	kim	kin	kio	kip	kir
	kof	kog	koh	koa	kok	kol	39	kom	kon	koy	kop	kor	kos
	kug	kuh	kua	kuk	kul	kum	40	kun	kuo	kup	kur	kus	kut
	kyh	kya	kyk	kyl	kym	kyn	41	kyo	kyp	kyr	kys	kyt	kyv
	lac	lad	laf	lag	lah	lai	42	lak	lal	lam	lan	lao	lap
	led	lef	leg	leh	lea	lek	43	lel	lem	len	leo	lep	ler
	lif	lig	lih	lia	lik	lil	44	lim	lin	lio	lip	lir	lis
	log	loh	loa	lok	lol	lom	45	lon	loy	lop	lor	los	lot
	luh	lua	luk	lul	lum	lun	46	luo	lup	lur	lus	lut	luv
	lya	lyk	lyl	lym	lyn	lyo	47	lyp	lyr	lys	lyt	lyv	lyw
	mad	maf	mag	mah	mai	mak	48	mal	mam	man	mao	map	mar
	mef	meg	meh	mea	mek	mel	49	mem	men	meo	mep	mer	mes

B

First 2 of the 5 letters:	ax	ec	ig	ok	uw	ca	di	go	lu	ny
	20	21	22	23	24	25	26	27	28	29

last 3 of the 5 letters:

1st	2nd	3rd	4th	5th	6th		7th	8th	9th	10th	11th	12th class
mig	mih	mia	mik	mil	mim	50	min	mio	mip	mir	mis	mit
moh	moa	mok	mol	mom	mon	51	moy	mop	mor	mos	mot	mov
mua	muk	mul	mum	mun	muo	52	mup	mur	mus	mut	muv	muw
myk	myl	mym	myn	myo	myp	53	myr	mys	myt	myv	myw	mye
naf	nag	nah	nai	nak	nal	54	nam	nan	nao	nap	nar	nas
neg	neh	nea	nek	nel	nem	55	nen	neo	nep	ner	nes	net
nih	nia	nik	nil	nim	nin	56	nio	nip	nir	nis	nit	niv
noa	nok	nol	nom	non	noy	57	nop	nor	nos	not	nov	now
nuk	nul	num	nun	nuo	nup	58	nur	nus	nut	nuv	nuw	nue
nyl	nym	nyn	nyo	nyp	nyr	59	nys	nyt	nyv	nyw	nye	nyz
pah	pai	pak	pal	pam	pan	60	pao	pag	par	pas	pat	pav
pea	pek	pel	pem	pen	peo	61	pep	per	pes	pet	pev	pey
pik	pil	pim	pin	pio	pip	62	pir	pis	pit	piv	piw	pie
pol	pom	pon	poy	pop	por	63	pos	pot	pov	pow	poe	poz
pum	pun	puo	pup	pur	pus	64	put	puv	puw	pue	puz	pub
pyn	pyo	pyp	pyr	pys	pyt	65	pyv	pyw	pye	pyz	pyb	pyc
rai	rak	ral	ram	ran	rao	66	rap	rar	ras	rat	rav	raw
rek	rel	rem	ren	reo	rep	67	rer	res	ret	rev	rey	reu
ril	rim	rin	rio	rip	rir	68	ris	rit	riv	riw	rie	riz
rom	ron	roy	rop	ror	ros	69	rot	rov	row	roe	roz	rob
run	ruo	rup	rur	rus	rut	70	ruv	ruw	rue	ruz	rub	ruc
ryo	ryp	ryr	rys	ryt	ryv	71	ryw	rye	ryz	ryb	ryc	ryd
sak	sal	sam	san	sao	sap	72	sar	sas	sat	sav	saw	sae
sel	sem	sen	seo	sep	ser	73	ses	set	sev	sey	seu	sez
sim	sin	sio	sip	sir	sis	74	sit	siv	siw	sie	siz	sib
son	soy	sop	sor	sos	sot	75	sov	sow	soe	soz	sob	soc
suo	sup	sur	sus	sut	suv	76	suw	sue	suz	sub	suc	sud
syp	syr	sys	syt	syv	syw	77	sye	syz	syb	syc	syd	syf
tal	tam	tan	tao	tap	tar	78	tas	tat	tav	taw	tae	taz
tem	ten	teo	tep	ter	tes	79	tet	tev	tey	teu	tez	teb
tin	tio	tip	tir	tis	tit	80	tiv	tiw	tie	tiz	tib	tic
toy	top	tor	tos	tot	tov	81	tow	toe	toz	tob	toc	tod
tup	tur	tus	tut	tuv	tuw	82	tue	tuz	tub	tuc	tud	tuf
tyr	tys	tyt	tyv	tyw	tye	83	tyz	tyb	tyc	tyd	tyf	tyg
vam	van	vao	vap	var	vas	84	vat	vav	vaw	vae	vaz	vab
ven	veo	vep	ver	ves	vet	85	vev	vey	veu	vez	veb	vec
vio	vip	vir	vis	vit	viv	86	viw	vie	viz	vib	vic	vid
vop	vor	vos	vot	vov	vow	87	voe	voz	vob	voc	vod	vof
vur	vus	vut	vuv	vuw	vue	88	vuz	vub	vuc	vud	vuf	vug
vys	vyt	vyv	vyw	vye	vyz	89	vyb	vyc	vyd	vyf	vyg	vyh
wan	wao	wap	war	was	wat	90	wav	waw	wae	waz	wab	wac
weo	wep	wer	wes	wet	wev	91	wey	weu	wez	web	wec	wed
wip	wir	wis	wit	wiv	woe	92	wie	wiz	wib	wic	wid	wif
wor	wos	wot	wov	wue	wuz	93	woz	wob	woc	wod	wof	wog
zap	zar	zas	zat	zav	zaw	94	zae	zaz	zab	zac	zad	zaf
zer	zes	zet	zev	zey	zeu	95	zez	zeb	zec	zed	zef	zeg
zis	zit	ziv	ziw	zie	ziz	96	zib	zic	zid	zif	zig	zih
zot	zov	zow	zoe	zoz	zob	97	zoc	zod	zof	zog	zoh	zoa
zuv	zuw	zue	zuz	zub	zuc	98	zud	zuf	zug	zuh	zua	zuk
zyw	zye	zyz	zyb	zyc	zyd	99	zyf	zyg	zyh	zya	zyk	zyl

B
20

B

First 2. of the 5 letters	as	eg	if	ol	un	be	wi	zo	su	py
	30	31	32	33	34	35	36	37	38	39

last 3 of the 5 letters:

	1st	2nd	3rd	4th	5th	6th		7th	8th	9th	10th	11th	12th class
B 30	bas	bat	bav	baw	bae	baz	**00**	bab	bac	bad	baf	bag	bah
	bet	bev	bey	beu	bez	beb	**01**	bec	bed	bef	beg	beh	bea
	biv	biw	bie	biz	bib	bic	**02**	bid	bif	big	bih	bia	bik
	bow	boe	boz	bob	boc	bod	**03**	bof	bog	boh	boa	bok	bol
	bue	buz	bub	buc	bud	buf	**04**	bug	buh	bua	buk	bul	bum
	byz	byb	byc	byd	byf	byg	**05**	byh	bya	byk	byl	bym	byn
	cat	cav	caw	cae	caz	cab	**06**	cac	cad	caf	cag	cah	cai
	cev	cey	ceu	cez	ceb	cec	**07**	ced	cef	ceg	ceh	cea	cek
	ciw	cie	ciz	cib	cic	cid	**08**	cif	cig	cih	cia	cik	cil
	coe	coz	cob	coc	cod	cof	**09**	cog	coh	coa	cok	col	com
	cuz	cub	cuc	cud	cuf	cug	**10**	cuh	cua	cuk	cul	cum	cun
	cyb	cyc	cyd	cyf	cyg	cyh	**11**	cya	cyk	cyl	cym	cyn	cyo
	dav	daw	dae	daz	dab	dac	**12**	dad	daf	dag	dah	dai	dak
	dey	deu	dez	deb	dec	ded	**13**	def	deg	deh	dea	dek	del
	die	diz	dib	dic	did	dif	**14**	dig	dih	dia	dik	dil	dim
	doz	dob	doc	dod	dof	dog	**15**	doh	doa	dok	dol	dom	don
	dub	duc	dud	duf	dug	duh	**16**	dua	duk	dul	dum	dun	duo
	dyc	dyd	dyf	dyg	dyh	dya	**17**	dyk	dyl	dym	dyn	dyo	dyp
	faw	fae	faz	fab	fac	fad	**18**	faf	fag	fah	fai	fak	fal
	feu	fez	feb	fec	fed	fef	**19**	feg	feh	fea	fek	fel	fem
	fiz	fib	fic	fid	fif	fig	**20**	fih	fia	fik	fil	fim	fin
	fob	foc	fod	fof	fog	foh	**21**	foa	fok	fol	fom	fon	foy
	fuc	fud	fuf	fug	fuh	fua	**22**	fuk	ful	fum	fun	fuo	fup
	fyd	fyf	fyg	fyh	fya	fyk	**23**	fyl	fym	fyn	fyo	fyp	fyr
	gae	gaz	gab	gac	gad	gaf	**24**	gag	gah	gai	gak	gal	gam
	gez	geb	gec	ged	gef	geg	**25**	geh	gea	gek	gel	gem	gen
	gib	gic	gid	gif	gig	gih	**26**	gia	gik	gil	gim	gin	gio
	goc	god	gof	gog	goh	goa	**27**	gok	gol	gom	gon	goy	gop
	gud	guf	gug	guh	gua	guk	**28**	gul	gum	gun	guo	gup	gur
	gyf	gyg	gyh	gya	gyk	gyl	**29**	gym	gyn	gyo	gyp	gyr	gys
	haz	hab	hac	had	haf	hag	**30**	hau	hai	hak	hal	ham	han
	heb	hec	hed	hef	heg	hei	**31**	hea	hek	hel	hem	hen	heo
	hic	hid	hif	hig	hiu	hia	**32**	hik	hil	him	hin	hio	hip
	hod	hof	hog	hoi	hoa	hok	**33**	hol	hom	hon	hoy	hop	hor
	huf	hug	hui	hua	huk	hul	**34**	hum	hun	huo	hup	hur	hus
	hyg	hyu	hya	hyk	hyl	hym	**35**	hyn	hyo	hyp	hyr	hys	hyt
	kac	kad	kaf	kag	kah	kai	**36**	kak	kal	kam	kan	kao	kap
	ked	kef	keg	keh	kea	kek	**37**	kel	kem	ken	keo	kep	ker
	kif	kig	kih	kia	kik	kil	**38**	kim	kin	kio	kip	kir	kis
	kog	koh	koa	kok	kol	kom	**39**	kon	koy	kop	kor	kos	kot
	kuh	kua	kuk	kul	kum	kun	**40**	kuo	kup	kur	kus	kut	kuv
	kya	kyk	kyl	kym	kyn	kyo	**41**	kyp	kyr	kys	kyt	kyv	kyw
	lad	laf	lag	lah	lai	lak	**42**	lal	lam	lan	lao	lap	lar
	lef	leg	leh	lea	lek	lel	**43**	lem	len	leo	lep	ler	les
	lig	lih	lia	lik	lil	lim	**44**	lin	lio	lip	lir	lis	lit
	loh	loa	lok	lol	lom	lon	**45**	loy	lop	lor	los	lot	lov
	lua	luk	lul	lum	lun	luo	**46**	lup	lur	lus	lut	luv	luw
	lyk	lyl	lym	lyn	lyo	lyp	**47**	lyr	lys	lyt	lyv	lyw	lye
	maf	mag	mah	mai	mak	mal	**48**	mam	man	mao	map	mar	mas
	meg	meh	mea	mek	mel	mem	**49**	men	meo	mep	mer	mes	met

B

First 2 of the 5 letters:	as	eg	if	ol	un	be	wi	zo	su	py
last 3 of the 5 letters:	30	31	32	33	34	35	36	37	38	39

B 30

1st	2nd	3rd	4th	5th	6th		7th	8th	9th	10th	11th	12th class
mih	mia	mik	mil	mim	min	**50**	mio	mip	mir	mis	mit	miv
moa	mok	mol	mom	mon	moy	**51**	mop	mor	mos	mot	mov	mow
muk	mul	mum	mun	muo	mup	**52**	mur	mus	mut	muv	muw	mue
myl	mym	myn	myo	myp	myr	**53**	mys	myt	myv	myw	mye	myz
nag	nah	nai	nak	nal	nam	**54**	nan	nao	nap	nar	nas	nat
neh	nea	nek	nel	nem	nen	**55**	neo	nep	ner	nes	net	nev
nia	nik	nil	nim	nin	nio	**56**	nip	nir	nis	nit	niv	niw
nok	nol	nom	non	noy	nop	**57**	nor	nos	not	nov	now	noe
nul	num	nun	nuo	nup	nur	**58**	nus	nut	nuv	nuw	nue	nuz
nym	nyn	nyo	nyp	nyr	nys	**59**	nyt	nyv	nyw	nye	nyz	nyb
pai	pak	pal	pam	pan	pao	**60**	pap	par	pas	pat	pav	paw
pek	pel	pem	pen	peo	pep	**61**	per	pes	pet	pev	pey	peu
pil	pim	pin	pio	pip	pir	**62**	pis	pit	piv	piw	pie	piz
pom	pon	poy	pop	por	pos	**63**	pot	pov	pow	poe	poz	pob
pun	puo	pup	pur	pus	put	**64**	puv	puw	pue	puz	pub	puc
pyo	pyp	pyr	pys	pyt	pyv	**65**	pyw	pye	pyz	pyb	pyc	pyd
rak	ral	ram	ran	rao	rap	**66**	rar	ras	rat	rav	raw	rae
rel	rem	ren	reo	rep	rer	**67**	res	ret	rev	rey	reu	rez
rim	rin	rio	rip	rir	ris	**68**	rit	riv	riw	rie	riz	rib
ron	roy	rop	ror	ros	rot	**69**	rov	row	roe	roz	rob	roc
ruo	rup	rur	rus	rut	ruv	**70**	ruw	rue	ruz	rub	ruc	rud
ryp	ryr	rys	ryt	ryv	ryw	**71**	rye	ryz	ryb	ryc	ryd	ryf
sal	sam	san	sao	sap	sar	**72**	sas	sat	sav	saw	sae	saz
sem	sen	seo	sep	ser	ses	**73**	set	sev	sey	seu	sez	seb
sin	sio	sip	sir	sis	sit	**74**	siv	siw	sie	siz	sib	sic
soy	sop	sor	sos	sot	sov	**75**	sow	soe	soz	sob	soc	sod
sup	sur	sus	sut	suv	suw	**76**	sue	suz	sub	suc	sud	suf
syr	sys	syt	syv	syw	sye	**77**	syz	syb	syc	syd	syf	syg
tam	tan	tao	tap	tar	tas	**78**	tat	tav	taw	tae	taz	tab
ten	teo	tep	ter	tes	tet	**79**	tev	tey	teu	tez	teb	tec
tio	tip	tir	tis	tit	tiv	**80**	tiw	tie	tiz	tib	tic	tid
top	tor	tos	tot	tov	tow	**81**	toe	toz	tob	toc	tod	tof
tur	tus	tut	tuv	tuw	tue	**82**	tuz	tub	tuc	tud	tuf	tug
tys	tyt	tyv	tyw	tye	tyz	**83**	tyb	tyc	tyd	tyf	tyg	tyh
van	vao	vap	var	vas	vat	**84**	vav	vaw	vae	vaz	vab	vac
veo	vep	ver	ves	vet	vev	**85**	vey	veu	vez	veb	vec	ved
vip	vir	vis	vit	viv	viw	**86**	vie	viz	vib	vic	vid	vif
vor	vos	vot	vov	vow	voe	**87**	voz	vob	voc	vod	vof	vog
vus	vut	vuv	vuw	vue	vuz	**88**	vub	vuc	vud	vuf	vug	vuh
vyt	vyv	vyw	vye	vyz	vyb	**89**	vyc	vyd	vyf	vyg	vyh	vya
wao	wap	war	was	wat	wav	**90**	waw	wae	waz	wab	wac	wad
wep	wer	wes	wet	wev	wey	**91**	weu	wez	web	wec	wed	wef
wir	wis	wit	wiv	woe	wie	**92**	wiz	wib	wic	wid	wif	wig
wos	wot	wov	wue	wuz	woz	**93**	wob	woc	wod	wof	wog	woh
zar	zas	zat	zav	zaw	zae	**94**	zaz	zab	zac	zad	zaf	zag
zes	zet	zev	zey	zeu	zez	**95**	zeb	zec	zed	zef	zeg	zeh
zit	ziv	ziw	zie	ziz	zib	**96**	zic	zid	zif	zig	zih	zia
zov	zow	zoe	zoz	zob	zoc	**97**	zod	zof	zog	zoh	zoa	zok
zuw	zue	zuz	zub	zuc	zud	**98**	zuf	zug	zuh	zua	zuk	zul
zye	zyz	zyb	zyc	zyd	zyf	**99**	zyg	zyh	zya	zyk	zyl	zym

B

First 2 of the 5 letters:

ar	ef	ic	os	ga	de	pi	mo	nu	cy
40	41	42	43	44	45	46	47	48	49

last 3 of the 5 letters:

	1st	2nd	3rd	4th	5th	6th		7th	8th	9th	10th	11th	12th class
B	bat	bav	baw	bae	baz	bab	00	bac	bad	baf	bag	bah	bai
	bev	bey	beu	bez	beb	bec	01	bed	bef	beg	beh	bea	bek
40	biw	bie	biz	bib	bic	bid	02	bif	big	bih	bia	bik	bil
	boe	boz	bob	boc	bod	bof	03	bog	boh	boa	bok	bol	bom
	buz	bub	buc	bud	buf	bug	04	buh	bua	buk	bul	bum	bun
	byb	byc	byd	byf	byg	byh	05	bya	byk	byl	bym	byn	byo
	cav	caw	cae	caz	cab	cac	06	cad	caf	cag	cah	cai	cak
	cey	ceu	cez	ceb	cec	ced	07	cef	ceg	ceh	cea	cek	cel
	cie	ciz	cib	cic	cid	cif	08	cig	cih	cia	cik	cil	cim
	coz	cob	coc	cod	cof	cog	09	coh	coa	cok	col	com	con
	cub	cuc	cud	cuf	cug	cuh	10	cua	cuk	cul	cum	cun	cuo
	cyc	cyd	cyf	cyg	cyh	cya	11	cyk	cyl	cym	cyn	cyo	cyp
	daw	dae	daz	dab	dac	dad	12	daf	dag	dah	dai	dak	dal
	deu	dez	deb	dec	ded	def	13	deg	deh	dea	dek	del	dem
	diz	dib	dic	did	dif	dig	14	dih	dia	dik	dil	dim	din
	dob	doc	dod	dof	dog	doh	15	doa	dok	dol	dom	don	doy
	duc	dud	duf	dug	duh	dua	16	duk	dul	dum	dun	duo	dup
	dyd	dyf	dyg	dyh	dya	dyk	17	dyl	dym	dyn	dyo	dyp	dyr
	fae	faz	fab	fac	fad	faf	18	fag	fah	fai	fak	fal	fam
	fez	feb	fec	fed	fef	feg	19	feh	fea	fek	fel	fem	fen
	fib	fic	fid	fif	fig	fih	20	fia	fik	fil	fim	fin	fio
	foc	fod	fof	fog	foh	foa	21	fok	fol	fom	fon	foy	fop
	fud	fuf	fug	fuh	fua	fuk	22	ful	fum	fun	fuo	fup	fur
	fyf	fyg	fyh	fya	fyk	fyl	23	fym	fyn	fyo	fyp	fyr	fys
	gaz	gab	gac	gad	gaf	gag	24	gah	gai	gak	gal	gam	gan
	geb	gec	ged	gef	geg	geh	25	gea	gek	gel	gem	gen	geo
	gic	gid	gif	gig	gih	gia	26	gik	gil	gim	gin	gio	gip
	god	gof	gog	goh	goa	gok	27	gol	gom	gon	goy	gop	gor
	guf	gug	guh	gua	guk	gul	28	gum	gun	guo	gup	gur	gus
	gyg	gyh	gya	gyk	gyl	gym	29	gyn	gyo	gyp	gyr	gys	gyt
	hab	hac	had	haf	hag	hau	30	hai	hak	hal	ham	han	hao
	hec	hed	hef	heg	hei	hea	31	hek	hel	hem	hen	heo	hep
	hid	hif	hig	hiu	hia	hik	32	hil	him	hin	hio	hip	hir
	hof	hog	hoi	hoa	hok	hol	33	hom	hon	hoy	hop	hor	hos
	hug	hui	hua	huk	hul	hum	34	hun	huo	hup	hur	hus	hut
	hyu	hya	hyk	hyl	hym	hyn	35	hyo	hyp	hyr	hys	hyt	hyv
	kad	kaf	kag	kah	kai	kak	36	kal	kam	kan	kao	kap	kar
	kef	keg	keh	kea	kek	kel	37	kem	ken	keo	kep	ker	kes
	kig	kih	kia	kik	kil	kim	38	kin	kio	kip	kir	kis	kit
	koh	koa	kok	kol	kom	kon	39	koy	kop	kor	kos	kot	kov
	kua	kuk	kul	kum	kun	kuo	40	kup	kur	kus	kut	kuv	kuw
	kyk	kyl	kym	kyn	kyo	kyp	41	kyr	kys	kyt	kyv	kyw	kye
	laf	lag	lah	lai	lak	lal	42	lam	lan	lao	lap	lar	las
	leg	leh	lea	lek	lel	lem	43	len	leo	lep	ler	les	let
	lih	lia	lik	lil	lim	lin	44	lio	lip	lir	lis	lit	liv
	loa	lok	lol	lom	lon	loy	45	lop	lor	los	lot	lov	low
	luk	lul	lum	lun	luo	lup	46	lur	lus	lut	luv	luw	lue
	lyl	lym	lyn	lyo	lyp	lyr	47	lys	lyt	lyv	lyw	lye	lyz
	mag	mah	mai	mak	mal	mam	48	man	mao	map	mar	mas	mat
	meh	mea	mek	mel	mem	men	49	meo	mep	mer	mes	met	mev

B

First 2 of the 5 letters:

ar	ef	ic	os	ga	de	pi	mo	nu	cy
40	41	42	43	44	45	46	47	48	49

last 3 of the 5 letters:

1st	2nd	3rd	4th	5th	6th		7th	8th	9th	10th	11th	12th class
mia	mik	mil	mim	min	mio	50	mip	mir	mis	mit	miv	miw
mok	mol	mom	mon	moy	mop	51	mor	mos	mot	mov	mow	moe
mul	mum	mun	muo	mup	mur	52	mus	mut	muv	muw	mue	muz
mym	myn	myo	myp	myr	mys	53	myt	myv	myw	mye	myz	myb
nah	nai	nak	nal	nam	nan	54	nao	nap	nar	nas	nat	nav
nea	nek	nel	nem	nen	neo	55	nep	ner	nes	net	nev	ney
nik	nil	nim	nin	nio	nip	56	nir	nis	nit	niv	niw	nie
nol	nom	non	noy	nop	nor	57	nos	not	nov	now	noe	noz
num	nun	nuo	nup	nur	nus	58	nut	nuv	nuw	nue	nuz	nub
nyn	nyo	nyp	nyr	nys	nyt	59	nyv	nyw	nye	nyz	nyb	nyc
pak	pal	pam	pan	pao	pap	60	par	pas	pat	pav	paw	pae
pel	pem	pen	peo	pep	per	61	pes	pet	pev	pey	peu	pez
pim	pin	pio	pip	pir	pis	62	pit	piv	piw	pie	piz	pib
pon	poy	pop	por	pos	pot	63	pov	pow	poe	poz	pob	poc
puo	pup	pur	pus	put	puv	64	puw	pue	puz	pub	puc	pud
pyp	pyr	pys	pyt	pyv	pyw	65	pye	pyz	pyb	pyc	pyd	pyf
ral	ram	ran	rao	rap	rar	66	ras	rat	rav	raw	rae	raz
rem	ren	reo	rep	rer	res	67	ret	rev	rey	reu	rez	reb
rin	rio	rip	rir	ris	rit	68	riv	riw	rie	riz	rib	ric
roy	rop	ror	ros	rot	rov	69	row	roe	roz	rob	roc	rod
rup	rur	rus	rut	ruv	ruw	70	rue	ruz	rub	ruc	rud	ruf
ryr	rys	ryt	ryv	ryw	rye	71	ryz	ryb	ryc	ryd	ryf	ryg
sam	san	sao	sap	sar	sas	72	sat	sav	saw	sae	saz	sab
sen	seo	sep	ser	ses	set	73	sev	sey	seu	sez	seb	sec
sio	sip	sir	sis	sit	siv	74	siw	sie	siz	sib	sic	sid
sop	sor	sos	sot	sov	sow	75	soe	soz	sob	soc	sod	sof
sur	sus	sut	suv	suw	sue	76	suz	sub	suc	sud	suf	sug
sys	syt	syv	syw	sye	syz	77	syb	syc	syd	syf	syg	syh
tan	tao	tap	tar	tas	tat	78	tav	taw	tae	taz	tab	tac
teo	tep	ter	tes	tet	tev	79	tey	teu	tez	teb	tec	ted
tip	tir	tis	tit	tiv	tiw	80	tie	tiz	tib	tic	tid	tif
tor	tos	tot	tov	tow	toe	81	toz	tob	toc	tod	tof	tog
tus	tut	tuv	tuw	tue	tuz	82	tub	tuc	tud	tuf	tug	tuh
tyt	tyv	tyw	tye	tyz	tyb	83	tyc	tyd	tyf	tyg	tyh	tya
vao	vap	var	vas	vat	vav	84	vaw	vae	vaz	vab	vac	vad
vep	ver	ves	vet	vev	vey	85	veu	vez	veb	vec	ved	vef
vir	vis	vit	viv	viw	vie	86	viz	vib	vic	vid	vif	vig
vos	vot	vov	vow	voe	voz	87	vob	voc	vod	vof	vog	voh
vut	vuv	vuw	vue	vuz	vub	88	vuc	vud	vuf	vug	vuh	vua
vyv	vyw	vye	vyz	vyb	vyc	89	vyd	vyf	vyg	vyh	vya	vyk
wap	war	was	wat	wav	waw	90	wae	waz	wab	wac	wad	waf
wer	wes	wet	wev	wey	weu	91	wez	web	wec	wed	wef	weg
wis	wit	wiv	woe	wie	wiz	92	wib	wic	wid	wif	wig	wih
wot	wov	wue	wuz	woz	wob	93	woc	wod	wof	wog	woh	woa
zas	zat	zav	zaw	zae	zaz	94	zab	zac	zad	zaf	zag	zah
zet	zev	zey	zeu	zez	zeb	95	zec	zed	zef	zeg	zeh	zea
ziv	ziw	zie	ziz	zib	zic	96	zid	zif	zig	zih	zia	zik
zow	zoe	zoz	zob	zoc	zod	97	zof	zog	zoh	zoa	zok	zol
zue	zuz	zub	zuc	zud	zuf	98	zug	zuh	zua	zuk	zul	zum
zyz	zyb	zyc	zyd	zyf	zyg	99	zyh	zya	zyk	zyl	zym	zyn

B
40

B

First 2 of the 5 letters:	at	en	iw	oz	ba	pe	vi	ko	mu	ly
last 3 of the 5 letters:	50	51	52	53	54	55	56	57	58	59

B 50

1st	2nd	3rd	4th	5th	6th		7th	8th	9th	10th	11th	12th class
bav	baw	bae	baz	bab	bac	00	bad	baf	bag	bah	bai	bak
bey	beu	bez	beb	bec	bed	01	bef	beg	beh	bea	bek	bel
bie	biz	bib	bic	bid	bif	02	big	bih	bia	bik	bil	bim
boz	bob	boc	bod	bof	bog	03	boh	boa	bok	bol	bom	bon
bub	buc	bud	buf	bug	buh	04	bua	buk	bul	bum	bun	buo
byc	byd	byf	byg	byh	bya	05	byk	byl	bym	byn	byo	byp
caw	cae	caz	cab	cac	cad	06	caf	cag	cah	cai	cak	cal
ceu	cez	ceb	cec	ced	cef	07	ceg	ceh	cea	cek	cel	cem
ciz	cib	cic	cid	cif	cig	08	cih	cia	cik	cil	cim	cin
cob	coc	cod	cof	cog	coh	09	coa	cok	col	com	con	coy
cuc	cud	cuf	cug	cuh	cua	10	cuk	cul	cum	cun	cuo	cup
cyd	cyf	cyg	cyh	cya	cyk	11	cyl	cym	cyn	cyo	cyp	cyr
dae	daz	dab	dac	dad	daf	12	dag	dah	dai	dak	dal	dam
dez	deb	dec	ded	def	deg	13	deh	dea	dek	del	dem	den
dib	dic	did	dif	dig	dih	14	dia	dik	dil	dim	din	dio
doc	dod	dof	dog	doh	doa	15	dok	dol	dom	don	doy	dop
dud	duf	dug	duh	dua	duk	16	dul	dum	dun	duo	dup	dur
dyf	dyg	dyh	dya	dyk	dyl	17	dym	dyn	dyo	dyp	dyr	dys
faz	fab	fac	fad	faf	fag	18	fah	fai	fak	fal	fam	fan
feb	fec	fed	fef	feg	feh	19	fea	fek	fel	fem	fen	feo
fic	fid	fif	fig	fih	fia	20	fik	fil	fim	fin	fio	fip
fod	fof	fog	foh	foa	fok	21	fol	fom	fon	foy	fop	for
fuf	fug	fuh	fua	fuk	ful	22	fum	fun	fuo	fup	fur	fus
fyg	fyh	fya	fyk	fyl	fym	23	fyn	fyo	fyp	fyr	fys	fyt
gab	gac	gad	gaf	gag	gah	24	gai	gak	gal	gam	gan	gao
gec	ged	gef	geg	geh	gea	25	gek	gel	gem	gen	geo	gep
gid	gif	gig	gih	gia	gik	26	gil	gim	gin	gio	gip	gir
gof	gog	goh	goa	gok	gol	27	gom	gon	goy	gop	gor	gos
gug	guh	gua	guk	gul	gum	28	gun	guo	gup	gur	gus	gut
gyh	gya	gyk	gyl	gym	gyn	29	gyo	gyp	gyr	gys	gyt	gyv
hac	had	haf	hag	hau	hai	30	hak	hal	ham	han	hao	hap
hed	hef	heg	hei	hea	hek	31	hel	hem	hen	heo	hep	her
hif	hig	hiu	hia	hik	hil	32	him	hin	hio	hip	hir	his
hog	hoi	hoa	hok	hol	hom	33	hon	hoy	hop	hor	hos	hot
hui	hua	huk	hul	hum	hun	34	huo	hup	hur	hus	hut	huv
hya	hyk	hyl	hym	hyn	hyo	35	hyp	hyr	hys	hyt	hyv	hyw
kaf	kag	kah	kai	kak	kal	36	kam	kan	kao	kap	kar	kas
keg	keh	kea	kek	kel	kem	37	ken	keo	kep	ker	kes	ket
kih	kia	kik	kil	kim	kin	38	kio	kip	kir	kis	kit	kiv
koa	kok	kol	kom	kon	koy	39	kop	kor	kos	kot	kov	kow
kuk	kul	kum	kun	kuo	kup	40	kur	kus	kut	kuv	kuw	kue
kyl	kym	kyn	kyo	kyp	kyr	41	kys	kyt	kyv	kyw	kye	kyz
lag	lah	lai	lak	lal	lam	42	lan	lao	lap	lar	las	lat
leh	lea	lek	lel	lem	len	43	leo	lep	ler	les	let	lev
lia	lik	lil	lim	lin	lio	44	lip	lir	lis	lit	liv	liw
lok	lol	lom	lon	loy	lop	45	lor	los	lot	lov	low	loe
lul	lum	lun	luo	lup	lur	46	lus	lut	luv	luw	lue	luz
lym	lyn	lyo	lyp	lyr	lys	47	lyt	lyv	lyw	lye	lyz	lyb
mah	mai	mak	mal	mam	man	48	mao	map	mar	mas	mat	mav
mea	mek	mel	mem	men	meo	49	mep	mer	mes	met	mev	mey

B

First 2 of the 5 letters: { at 50 | en 51 | iw 52 | oz 53 | ba 54 | pe 55 | vi 56 | ko 57 | mu 58 | ly 59

last 3 of the 5 letters:

1st	2nd	3rd	4th	5th	6th		7th	8th	9th	10th	11th	12th class
mik	mil	mim	min	mio	mip	50	mir	mis	mit	miv	miw	mie
mol	mom	mon	moy	mop	mor	51	mos	mot	mov	mow	moe	moz
mum	mun	muo	mup	mur	mus	52	mut	muv	muw	mue	muz	mub
myn	myo	myp	myr	mys	myt	53	myv	myw	mye	myz	myb	myc
nai	nak	nal	nam	nan	nao	54	nap	nar	nas	nat	nav	naw
nek	nel	nem	nen	neo	nep	55	ner	nes	net	nev	ney	neu
nil	nim	nin	nio	nip	nir	56	nis	nit	niv	niw	nie	niz
nom	non	noy	nop	nor	nos	57	not	nov	now	noe	noz	nob
nun	nuo	nup	nur	nus	nut	58	nuv	nuw	nue	nuz	nub	nuc
nyo	nyp	nyr	nys	nyt	nyv	59	nyw	nye	nyz	nyb	nyc	nyd
pal	pam	pan	pao	pap	par	60	pas	pat	pav	paw	pae	paz
pem	pen	peo	pep	per	pes	61	pet	pev	pey	peu	pez	peb
pin	pio	pip	pir	pis	pit	62	piv	piw	pie	piz	pib	pic
poy	pop	por	pos	pot	pov	63	pow	poe	poz	pob	poc	pod
pup	pur	pus	put	puv	puw	64	pue	puz	pub	puc	pud	puf
pyr	pys	pyt	pyv	pyw	pye	65	pyz	pyb	pyc	pyd	pyf	pyg
ram	ran	rao	rap	rar	ras	66	rat	rav	raw	rae	raz	rab
ren	reo	rep	rer	res	ret	67	rev	rey	reu	rez	reb	rec
rio	rip	rir	ris	rit	riv	68	riw	rie	riz	rib	ric	rid
rop	ror	ros	rot	rov	row	69	roe	roz	rob	roc	rod	rof
rur	rus	rut	ruv	ruw	rue	70	ruz	rub	ruc	rud	ruf	rug
rys	ryt	ryv	ryw	rye	ryz	71	ryb	ryc	ryd	ryf	ryg	ryh
san	sao	sap	sar	sas	sat	72	sav	saw	sae	saz	sab	sac
seo	sep	ser	ses	set	sev	73	sey	seu	sez	seb	sec	sed
sip	sir	sis	sit	siv	siw	74	sie	siz	sib	sic	sid	sif
sor	sos	sot	sov	sow	soe	75	soz	sob	soc	sod	sof	sog
sus	sut	suv	suw	sue	suz	76	sub	suc	sud	suf	sug	suh
syt	syv	syw	sye	syz	syb	77	syc	syd	syf	syg	syh	sya
tao	tap	tar	tas	tat	tav	78	taw	tae	taz	tab	tac	tad
tep	ter	tes	tet	tev	tey	79	teu	tez	teb	tec	ted	tef
tir	tis	tit	tiv	tiw	tie	80	tiz	tib	tic	tid	tif	tig
tos	tot	tov	tow	toe	toz	81	tob	toc	tod	tof	tog	toh
tut	tuv	tuw	tue	tuz	tub	82	tuc	tud	tuf	tug	tuh	tua
tyv	tyw	tye	tyz	tyb	tyc	83	tyd	tyf	tyg	tyh	tya	tyk
vap	var	vas	vat	vav	vaw	84	vae	vaz	vab	vac	vad	vaf
ver	ves	vet	vev	vey	veu	85	vez	veb	vec	ved	vef	veg
vis	vit	viv	viw	vie	viz	86	vib	vic	vid	vif	vig	vih
vot	vov	vow	voe	voz	vob	87	voc	vod	vof	vog	voh	voa
vuv	vuw	vue	vuz	vub	vuc	88	vud	vuf	vug	vuh	vua	vuk
vyw	vye	vyz	vyb	vyc	vyd	89	vyf	vyg	vyh	vya	vyk	vyl
war	was	wat	wav	waw	wae	90	waz	wab	wac	wad	waf	wag
wes	wet	wev	wey	weu	wez	91	web	wec	wed	wef	weg	weh
wit	wiv	woe	wie	wiz	wib	92	wic	wid	wif	wig	wih	wia
wov	wue	wuz	woz	wob	woc	93	wod	wof	wog	woh	woa	wok
zat	zav	zaw	zae	zaz	zab	94	zac	zad	zaf	zag	zah	zai
zev	zey	zeu	zez	zeb	zec	95	zed	zef	zeg	zeh	zea	zek
ziw	zie	ziz	zib	zic	zid	96	zif	zig	zih	zia	zik	zil
zoe	zoz	zob	zoc	zod	zof	97	zog	zoh	zoa	zok	zol	zom
zuz	zub	zuc	zud	zuf	zug	98	zuh	zua	zuk	zul	zum	zun
zyb	zyc	zyd	zyf	zyg	zyh	99	zya	zyk	zyl	zym	zyn	zyo

B
50

B

First 2 of the 5 letters { aw 60 | ek 61 | is 62 | om 63 | ub 64 | da 65 | we 66 | gi 67 | no 68 | zu 69

last 3 of the 5 letters:

	1st	2nd	3rd	4th	5th	6th		7th	8th	9th	10th	11th	12th class
B	baw	bae	baz	bab	bac	bad	00	baf	bag	bah	bai	bak	bal
	beu	bez	beb	bec	bed	bef	01	beg	beh	bea	bek	bel	bem
60	biz	bib	bic	bid	bif	big	02	bih	bia	bik	bil	bim	bin
	bob	boc	bod	bof	bog	boh	03	boa	bok	bol	bom	bon	boy
	buc	bud	buf	bug	buh	bua	04	buk	bul	bum	bun	buo	bup
	byd	byf	byg	byh	bya	byk	05	byl	bym	byn	byo	byp	byr
	cae	caz	cab	cac	cad	caf	06	cag	cah	cai	cak	cal	cam
	cez	ceb	cec	ced	cef	ceg	07	ceh	cea	cek	cel	cem	cen
	cib	cic	cid	cif	cig	cih	08	cia	cik	cil	cim	cin	cio
	coc	cod	cof	cog	coh	coa	09	cok	col	com	con	coy	cop
	cud	cuf	cug	cuh	cua	cuk	10	cul	cum	cun	cuo	cup	cur
	cyf	cyg	cyh	cya	cyk	cyl	11	cym	cyn	cyo	cyp	cyr	cys
	daz	dab	dac	dad	daf	dag	12	dah	dai	dak	dal	dam	dan
	deb	dec	ded	def	deg	deh	13	dea	dek	del	dem	den	deo
	dic	did	dif	dig	dih	dia	14	dik	dil	dim	din	dio	dip
	dod	dof	dog	doh	doa	dok	15	dol	dom	don	doy	dop	dor
	duf	dug	duh	dua	duk	dul	16	dum	dun	duo	dup	dur	dus
	dyg	dyh	dya	dyk	dyl	dym	17	dyn	dyo	dyp	dyr	dys	dyt
	fab	fac	fad	faf	fag	fah	18	fai	fak	fal	fam	fan	fao
	fec	fed	fef	feg	feh	fea	19	fek	fel	fem	fen	feo	fep
	fid	fif	fig	fih	fia	fik	20	fil	fim	fin	fio	fip	fir
	fof	fog	foh	foa	fok	fol	21	fom	fon	foy	fop	for	fos
	fug	fuh	fua	fuk	ful	fum	22	fun	fuo	fup	fur	fus	fut
	fyh	fya	fyk	fyl	fym	fyn	23	fyo	fyp	fyr	fys	fyt	fyv
	gac	gad	gaf	gag	gah	gai	24	gak	gal	gam	gan	gao	gap
	ged	gef	geg	geh	gea	gek	25	gel	gem	gen	geo	gep	ger
	gif	gig	gih	gia	gik	gil	26	gim	gin	gio	gip	gir	gis
	gog	goh	goa	gok	gol	gom	27	gon	goy	gop	gor	gos	got
	guh	gua	guk	gul	gum	gun	28	guo	gup	gur	gus	gut	guv
	gya	gyk	gyl	gym	gyn	gyo	29	gyp	gyr	gys	gyt	gyv	gyw
	had	haf	hag	hau	hai	hak	30	hal	ham	han	hao	hap	har
	hef	heg	hei	hea	hek	hel	31	hem	hen	heo	hep	her	hes
	hig	hiu	hia	hik	hil	him	32	hin	hio	hip	hir	his	hit
	hoi	hoa	hok	hol	hom	hon	33	hoy	hop	hor	hos	hot	hov
	hua	huk	hul	hum	hun	huo	34	hup	hur	hus	hut	huv	huw
	hyk	hyl	hym	hyn	hyo	hyp	35	hyr	hys	hyt	hyv	hyw	hye
	kag	kah	kai	kak	kal	kam	36	kan	kao	kap	kar	kas	kat
	keh	kea	kek	kel	kem	ken	37	keo	kep	ker	kes	ket	kev
	kia	kik	kil	kim	kin	kio	38	kip	kir	kis	kit	kiv	kiw
	kok	kol	kom	kon	koy	kop	39	kor	kos	kot	kov	kow	koe
	kul	kum	kun	kuo	kup	kur	40	kus	kut	kuv	kuw	kue	kuz
	kym	kyn	kyo	kyp	kyr	kys	41	kyt	kyv	kyw	kye	kyz	kyb
	lah	lai	lak	lal	lam	lan	42	lao	lap	lar	las	lat	lav
	lea	lek	lel	lem	len	leo	43	lep	ler	les	let	lev	ley
	lik	lil	lim	lin	lio	lip	44	lir	lis	lit	liv	liw	lie
	lol	lom	lon	loy	lop	lor	45	los	lot	lov	low	loe	loz
	lum	lun	luo	lup	lur	lus	46	lut	luv	luw	lue	luz	lub
	lyn	lyo	lyp	lyr	lys	lyt	47	lyv	lyw	lye	lyz	lyb	lyc
	mai	mak	mal	mam	man	mao	48	map	mar	mas	mat	mav	maw
	mek	mel	mem	men	meo	mep	49	mer	mes	met	mev	mey	meu

B

First 2 of the 5 letters:	aw	ek	is	om	ub	da	we	gi	no	zu
	60	61	62	63	64	65	66	67	68	69

last 3 of the 5 letters:

1st	2nd	3rd	4th	5th	6th		7th	8th	9th	10th	11th	12th class
mil	mim	min	mio	mip	mir	50	mis	mit	miv	miw	mie	miz
mom	mon	moy	mop	mor	mos	51	mot	mov	mow	moe	moz	mob
mun	muo	mup	mur	mus	mut	52	muv	muw	mue	muz	mub	muc
myo	myp	myr	mys	myt	myv	53	myw	mye	myz	myb	myc	myd
nak	nal	nam	nan	nao	nap	54	nar	nas	nat	nav	naw	nae
nel	nem	nen	neo	nep	ner	55	nes	net	nev	ney	neu	nez
nim	nin	nio	nip	nir	nis	56	nit	niv	niw	nie	niz	nib
non	noy	nop	nor	nos	not	57	nov	now	noe	noz	nob	noc
nuo	nup	nur	nus	nut	nuv	58	nuw	nue	nuz	nub	nuc	nud
nyp	nyr	nys	nyt	nyv	nyw	59	nye	nyz	nyb	nyc	nyd	nyf
pam	pan	pao	pag	par	pas	60	pat	pav	paw	pae	paz	pab
pen	peo	pep	per	pes	pet	61	pev	pey	peu	pez	peb	pec
pio	pip	pir	pis	pit	piv	62	piw	pie	piz	pib	pic	pid
pop	por	pos	pot	pov	pow	63	poe	poz	pob	poc	pod	pof
pur	pus	put	puv	puw	pue	64	puz	pub	puc	pud	puf	pug
pys	pyt	pyv	pyw	pye	pyz	65	pyb	pyc	pyd	pyf	pyg	pyh
ran	rao	rap	rar	ras	rat	66	rav	raw	rae	raz	rab	rac
reo	rep	rer	res	ret	rev	67	rey	reu	rez	reb	rec	red
rip	rir	ris	rit	riv	riw	68	rie	riz	rib	ric	rid	rif
ror	ros	rot	rov	row	roe	69	roz	rob	roc	rod	rof	rog
rus	rut	ruv	ruw	rue	ruz	70	rub	ruc	rud	ruf	rug	ruh
ryt	ryv	ryw	rye	ryz	ryb	71	ryc	ryd	ryf	ryg	ryh	rya
sao	sap	sar	sas	sat	sav	72	saw	sae	saz	sab	sac	sad
sep	ser	ses	set	sev	sey	73	seu	sez	seb	sec	sed	sef
sir	sis	sit	siv	siw	sie	74	siz	sib	sic	sid	sif	sig
sos	sot	sov	sow	soe	soz	75	sob	soc	sod	sof	sog	soh
sut	suv	suw	sue	suz	sub	76	suc	sud	suf	sug	suh	sua
syv	syw	sye	syz	syb	syc	77	syd	syf	syg	syh	sya	syk
tap	tar	tas	tat	tav	taw	78	tae	taz	tab	tac	tad	taf
ter	tes	tet	tev	tey	teu	79	tez	teb	tec	ted	tef	teg
tis	tit	tiv	tiw	tie	tiz	80	tib	tic	tid	tif	tig	tih
tot	tov	tow	toe	toz	tob	81	toc	tod	tof	tog	toh	toa
tuv	tuw	tue	tuz	tub	tuc	82	tud	tuf	tug	tuh	tua	tuk
tyw	tye	tyz	tyb	tyc	tyd	83	tyf	tyg	tyh	tya	tyk	tyl
var	vas	vat	vav	vaw	vae	84	vaz	vab	vac	vad	vaf	vag
ves	vet	vev	vey	veu	vez	85	veb	vec	ved	vef	veg	veh
vit	viv	viw	vie	viz	vib	86	vic	vid	vif	vig	vih	via
vov	vow	voe	voz	vob	voc	87	vod	vof	vog	voh	voa	vok
vuw	vue	vuz	vub	vuc	vud	88	vuf	vug	vuh	vua	vuk	vul
vye	vyz	vyb	vyc	vyd	vyf	89	vyg	vyh	vya	vyk	vyl	vym
was	wat	wav	waw	wae	waz	90	wab	wac	wad	waf	wag	wah
wet	wev	wey	weu	wez	web	91	wec	wed	wef	weg	weh	wea
wiv	woe	wie	wiz	wib	wic	92	wid	wif	wig	wih	wia	wik
wue	wuz	woz	wob	woc	wod	93	wof	wog	woh	woa	wok	wol
zav	zaw	zae	zaz	zab	zac	94	zad	zaf	zag	zah	zai	zak
zey	zeu	zez	zeb	zec	zed	95	zef	zeg	zeh	zea	zek	zel
zie	ziz	zib	zic	zid	zif	96	zig	zih	zia	zik	zil	zim
zoz	zob	zoc	zod	zof	zog	97	zoh	zoa	zok	zol	zom	zon
zub	zuc	zud	zuf	zug	zuh	98	zua	zuk	zul	zum	zun	zuo
zyc	zyd	zyf	zyg	zyh	zya	99	zyk	zyl	zym	zyn	zyo	zyp

B
60

B

First 2 of the 5 letters: av 70 · ep 71 · od 72 · uc 73 · fa 74 · he 75 · mi 76 · ro 77 · ju 78 · ty 79

	av	ep	od	uc	fa	he	mi	ro	ju	ty
First 2 of the 5 letters:	70	71	72	73	74	75	76	77	78	79

last 3 of the 5 letters:

	1st	2nd	3rd	4th	5th	6th		7th	8th	9th	10th	11th	12th class
B	bae	baz	bab	bac	bad	baf	00	bag	bah	bai	bak	bal	bam
	bez	beb	bec	bed	bef	beg	01	beh	bea	bek	bel	bem	ben
70	bib	bic	bid	bif	big	bih	02	bia	bik	bil	bim	bin	bio
	boc	bod	bof	bog	boh	boa	03	bok	bol	bom	bon	boy	bop
	bud	buf	bug	buh	bua	buk	04	bul	bum	bun	buo	bup	bur
	byf	byg	byh	bya	byk	byl	05	bym	byn	byo	byp	byr	bys
	caz	cab	cac	cad	caf	cag	06	cah	cai	cak	cal	cam	can
	ceb	cec	ced	cef	ceg	ceh	07	cea	cek	cel	cem	cen	ceo
	cic	cid	cif	cig	cih	cia	08	cik	cil	cim	cin	cio	cip
	cod	cof	cog	coh	coa	cok	09	col	com	con	coy	cop	cor
	cuf	cug	cuh	cua	cuk	cul	10	cum	cun	cuo	cup	cur	cus
	cyg	cyh	cya	cyk	cyl	cym	11	cyn	cyo	cyp	cyr	cys	cyt
	dab	dac	dad	daf	dag	dah	12	dai	dak	dal	dam	dan	dao
	dec	ded	def	deg	deh	dea	13	dek	del	dem	den	deo	dep
	did	dif	dig	dih	dia	dik	14	dil	dim	din	dio	dip	dir
	dof	dog	doh	doa	dok	dol	15	dom	don	doy	dop	dor	dos
	dug	duh	dua	duk	dul	dum	16	dun	duo	dup	dur	dus	dut
	dyh	dya	dyk	dyl	dym	dyn	17	dyo	dyp	dyr	dys	dyt	dyv
	fac	fad	faf	fag	fah	fai	18	fak	fal	fam	fan	fao	fap
	fed	fef	feg	feh	fea	fek	19	fel	fem	fen	feo	fep	fer
	fif	fig	fih	fia	fik	fil	20	fim	fin	fio	fip	fir	fis
	fog	foh	foa	fok	fol	fom	21	fon	foy	fop	for	fos	fot
	fuh	fua	fuk	ful	fum	fun	22	fuo	fup	fur	fus	fut	fuv
	fya	fyk	fyl	fym	fyn	fyo	23	fyp	fyr	fys	fyt	fyv	fyw
	gad	gaf	gag	gah	gai	gak	24	gal	gam	gan	gao	gap	gar
	gef	geg	geh	gea	gek	gel	25	gem	gen	geo	gep	ger	ges
	gig	gih	gia	gik	gil	gim	26	gin	gio	gip	gir	gis	git
	goh	goa	gok	gol	gom	gon	27	goy	gop	gor	gos	got	gov
	gua	guk	gul	gum	gun	guo	28	gup	gur	gus	gut	guv	guw
	gyk	gyl	gym	gyn	gyo	gyp	29	gyr	gys	gyt	gyv	gyw	gye
	haf	hag	hau	hai	hak	hal	30	ham	han	hao	hap	har	has
	heg	hei	hea	hek	hel	hem	31	hen	heo	hep	her	hes	het
	hiu	hia	hik	hil	him	hin	32	hio	hip	hir	his	hit	hiv
	hoa	hok	hol	hom	hon	hoy	33	hop	hor	hos	hot	hov	how
	huk	hul	hum	hun	huo	hup	34	hur	hus	hut	huv	huw	hue
	hyl	hym	hyn	hyo	hyp	hyr	35	hys	hyt	hyv	hyw	hye	hyz
	kah	kai	kak	kal	kam	kan	36	kao	kap	kar	kas	kat	kav
	kea	kek	kel	kem	ken	keo	37	kep	ker	kes	ket	kev	key
	kik	kil	kim	kin	kio	kip	38	kir	kis	kit	kiv	kiw	kie
	kol	kom	kon	koy	kop	kor	39	kos	kot	kov	kow	koe	koz
	kum	kun	kuo	kup	kur	kus	40	kut	kuv	kuw	kue	kuz	kub
	kyn	kyo	kyp	kyr	kys	kyt	41	kyv	kyw	kye	kyz	kyb	kyc
	lai	lak	lal	lam	lan	lao	42	lap	lar	las	lat	lav	law
	lek	lel	lem	len	leo	lep	43	ler	les	let	lev	ley	leu
	lil	lim	lin	lio	lip	lir	44	lis	lit	liv	liw	lie	liz
	lom	lon	loy	lop	lor	los	45	lot	lov	low	loe	loz	lob
	lun	luo	lup	lur	lus	lut	46	luv	luw	lue	luz	lub	luc
	lyo	lyp	lyr	lys	lyt	lyv	47	lyw	lye	lyz	lyb	lyc	lyd
	mak	mal	mam	man	mao	map	48	mar	mas	mat	mav	maw	mae
	mel	mem	men	meo	mep	mer	49	mes	met	mev	mey	meu	mez

B

First 2 of the 5 letters	av	ep	od	uc	fa	he	mi	ro	ju	ty
	70	71	72	73	74	75	76	77	78	79

last 3 of the 5 letters:

1st	2nd	3rd	4th	5th	6th		7th	8th	9th	10th	11th	12th class	
mim	min	mio	mip	mir	mis	**50**	mit	miv	miw	mie	miz	mib	**B**
mon	moy	mop	mor	mos	mot	**51**	mov	mow	moe	moz	mob	moc	
muo	mup	mur	mus	mut	muv	**52**	muw	mue	muz	mub	muc	mud	**70**
myp	myr	mys	myt	myv	myw	**53**	mye	myz	myb	myc	myd	myf	
nal	nam	nan	nao	nap	nar	**54**	nas	nat	nav	naw	nae	naz	
nem	nen	neo	nep	ner	nes	**55**	net	nev	ney	neu	nez	neb	
nin	nio	nip	nir	nis	nit	**56**	niv	niw	nie	niz	nib	nic	
noy	nop	nor	nos	not	nov	**57**	now	noe	noz	nob	noc	nod	
nup	nur	nus	nut	nuv	nuw	**58**	nue	nuz	nub	nuc	nud	nuf	
nyr	nys	nyt	nyv	nyw	nye	**59**	nyz	nyb	nyc	nyd	nyf	nyg	
pan	pao	pap	par	pas	pat	**60**	pav	paw	pae	paz	pab	pac	
peo	pep	per	pes	pet	pev	**61**	pey	peu	pez	peb	pec	ped	
pip	pir	pis	pit	piv	piw	**62**	pie	piz	pib	pic	pid	pif	
por	pos	pot	pov	pow	poe	**63**	poz	pob	poc	pod	pof	pog	
pus	put	puv	puw	pue	puz	**64**	pub	puc	pud	puf	pug	puh	
pyt	pyv	pyw	pye	pyz	pyb	**65**	pyc	pyd	pyf	pyg	pyh	pya	
rao	rap	rar	ras	rat	rav	**66**	raw	rae	raz	rab	rac	rad	
rep	rer	res	ret	rev	rey	**67**	reu	rez	reb	rec	red	ref	
rir	ris	rit	riv	riw	rie	**68**	riz	rib	ric	rid	rif	rig	
ros	rot	rov	row	roe	roz	**69**	rob	roc	rod	rof	rog	roh	
rut	ruv	ruw	rue	ruz	rub	**70**	ruc	rud	ruf	rug	ruh	rua	
ryv	ryw	rye	ryz	ryb	ryc	**71**	ryd	ryf	ryg	ryh	rya	ryk	
sap	sar	sas	sat	sav	saw	**72**	sae	saz	sab	sac	sad	saf	
ser	ses	set	sev	sey	seu	**73**	sez	seb	sec	sed	sef	seg	
sis	sit	siv	siw	sie	siz	**74**	sib	sic	sid	sif	sig	sih	
sot	sov	sow	soe	soz	sob	**75**	soc	sod	sof	sog	soh	soa	
suv	suw	sue	suz	sub	suc	**76**	sud	suf	sug	suh	sua	suk	
syw	sye	syz	syb	syc	syd	**77**	syf	syg	syh	sya	syk	syl	
tar	tas	tat	tav	taw	tae	**78**	taz	tab	tac	tad	taf	tag	
tes	tet	tev	tey	teu	tez	**79**	teb	tec	ted	tef	teg	teh	
tit	tiv	tiw	tie	tiz	tib	**80**	tic	tid	tif	tig	tih	tia	
tov	tow	toe	toz	tob	toc	**81**	tod	tof	tog	toh	toa	tok	
tuw	tue	tuz	tub	tuc	tud	**82**	tuf	tug	tuh	tua	tuk	tul	
tye	tyz	tyb	tyc	tyd	tyf	**83**	tyg	tyh	tya	tyk	tyl	tym	
vas	vat	vav	vaw	vae	vaz	**84**	vab	vac	vad	vaf	vag	vah	
vet	vev	vey	veu	vez	veb	**85**	vec	ved	vef	veg	veh	vea	
viv	viw	vie	viz	vib	vic	**86**	vid	vif	vig	vih	via	vik	
vow	voe	voz	vob	voc	vod	**87**	vof	vog	voh	voa	vok	vol	
vue	vuz	vub	vuc	vud	vuf	**88**	vug	vuh	vua	vuk	vul	vum	
vyz	vyb	vyc	vyd	vyf	vyg	**89**	vyh	vya	vyk	vyl	vym	vyn	
wat	wav	waw	wae	waz	wab	**90**	wac	wad	waf	wag	wah	wai	
wev	wey	weu	wez	web	wec	**91**	wed	wef	weg	weh	wea	wek	
woe	wie	wiz	wib	wic	wid	**92**	wif	wig	wih	wia	wik	wil	
wuz	woz	wob	woc	wod	wof	**93**	wog	woh	woa	wok	wol	wom	
zaw	zae	zaz	zab	zac	zad	**94**	zaf	zag	zah	zai	zak	zal	
zeu	zez	zeb	zec	zed	zef	**95**	zeg	zeh	zea	zek	zel	zem	
ziz	zib	zic	zid	zif	zig	**96**	zih	zia	zik	zil	zim	zin	
zob	zoc	zod	zof	zog	zoh	**97**	zoa	zok	zol	zom	zon	zoy	
zuc	zud	zuf	zug	zuh	zua	**98**	zuk	zul	zum	zun	zuo	zup	
zyd	zyf	zyg	zyh	zya	zyk	**99**	zyl	zym	zyn	zyo	zyp	zyr	

B

First 2 of the 5 letters:	ag	ix	op	uk	ka	te	si	co	bu	vy
last 3 of the 5 letters:	80	81	82	83	84	85	86	87	88	89

	1st	2nd	3rd	4th	5th	6th		7th	8th	9th	10th	11th	12th class
B	baz	bab	bac	bad	baf	bag	**00**	bah	bai	bak	bal	bam	ban
	beb	bec	bed	bef	beg	beh	**01**	bea	bek	bel	bem	ben	beo
80	bic	bid	bif	big	bih	bia	**02**	bik	bil	bim	bin	bio	bip
	bod	bof	bog	boh	boa	bok	**03**	bol	bom	bon	boy	bop	bor
	buf	bug	buh	bua	buk	bul	**04**	bum	bun	buo	bup	bur	bus
	byg	byh	bya	byk	byl	bym	**05**	byn	byo	byp	byr	bys	byt
	cab	cac	cad	caf	cag	cah	**06**	cai	cak	cal	cam	can	cao
	cec	ced	cef	ceg	ceh	cea	**07**	cek	cel	cem	cen	ceo	cep
	cid	cif	cig	cih	cia	cik	**08**	cil	cim	cin	cio	cip	cir
	cof	cog	coh	coa	cok	col	**09**	com	con	coy	cop	cor	cos
	cug	cuh	cua	cuk	cul	cum	**10**	cun	cuo	cup	cur	cus	cut
	cyh	cya	cyk	cyl	cym	cyn	**11**	cyo	cyp	cyr	cys	cyt	cyv
	dac	dad	daf	dag	dah	dai	**12**	dak	dal	dam	dan	dao	dap
	ded	def	deg	deh	dea	dek	**13**	del	dem	den	deo	dep	der
	dif	dig	dih	dia	dik	dil	**14**	dim	din	dio	dip	dir	dis
	dog	doh	doa	dok	dol	dom	**15**	don	doy	dop	dor	dos	dot
	duh	dua	duk	dul	dum	dun	**16**	duo	dup	dur	dus	dut	duv
	dya	dyk	dyl	dym	dyn	dyo	**17**	dyp	dyr	dys	dyt	dyv	dyw
	fad	faf	fag	fah	fai	fak	**18**	fal	fam	fan	fao	fap	far
	fef	feg	feh	fea	fek	fel	**19**	fem	fen	feo	fep	fer	fes
	fig	fih	fia	fik	fil	fim	**20**	fin	fio	fip	fir	fis	fit
	foh	foa	fok	fol	fom	fon	**21**	foy	fop	for	fos	fot	fov
	fua	fuk	ful	fum	fun	fuo	**22**	fup	fur	fus	fut	fuv	fuw
	fyk	fyl	fym	fyn	fyo	fyp	**23**	fyr	fys	fyt	fyv	fyw	fye
	gaf	gag	gah	gai	gak	gal	**24**	gam	gan	gao	gap	gar	gas
	geg	geh	gea	gek	gel	gem	**25**	gen	geo	gep	ger	ges	get
	gih	gia	gik	gil	gim	gin	**26**	gio	gip	gir	gis	git	giv
	goa	gok	gol	gom	gon	goy	**27**	gop	gor	gos	got	gov	gow
	guk	gul	gum	gun	guo	gup	**28**	gur	gus	gut	guv	guw	gue
	gyl	gym	gyn	gyo	gyp	gyr	**29**	gys	gyt	gyv	gyw	gye	gyz
	hag	hau	hai	hak	hal	ham	**30**	han	hao	hap	har	has	hat
	hei	hea	hek	hel	hem	hen	**31**	heo	hep	her	hes	het	hev
	hia	hik	hil	him	hin	hio	**32**	hip	hir	his	hit	hiv	hiw
	hok	hol	hom	hon	hoy	hop	**33**	hor	hos	hot	hov	how	hoe
	hul	hum	hun	huo	hup	hur	**34**	hus	hut	huv	huw	hue	huz
	hym	hyn	hyo	hyp	hyr	hys	**35**	hyt	hyv	hyw	hye	hyz	hyb
	kai	kak	kal	kam	kan	kao	**36**	kap	kar	kas	kat	kav	kaw
	kek	kel	kem	ken	keo	kep	**37**	ker	kes	ket	kev	key	keu
	kil	kim	kin	kio	kip	kir	**38**	kis	kit	kiv	kiw	kie	kiz
	kom	kon	koy	kop	kor	kos	**39**	kot	kov	kow	koe	koz	kob
	kun	kuo	kup	kur	kus	kut	**40**	kuv	kuw	kue	kuz	kub	kuc
	kyo	kyp	kyr	kys	kyt	kyv	**41**	kyw	kye	kyz	kyb	kyc	kyd
	lak	lal	lam	lan	lao	lap	**42**	lar	las	lat	lav	law	lae
	lel	lem	len	leo	lep	ler	**43**	les	let	lev	ley	leu	lez
	lim	lin	lio	lip	lir	lis	**44**	lit	liv	liw	lie	liz	lib
	lon	loy	lop	lor	los	lot	**45**	lov	low	loe	loz	lob	loc
	luo	lup	lur	lus	lut	luv	**46**	luw	lue	luz	lub	luc	lud
	lyp	lyr	lys	lyt	lyv	lyw	**47**	lye	lyz	lyb	lyc	lyd	lyf
	mal	mam	man	mao	map	mar	**48**	mas	mat	mav	maw	mae	maz
	mem	men	meo	mep	mer	mes	**49**	met	mev	mey	meu	mez	meb

B

First 2 of the 5 letters: { ag/80 ix/81 op/82 uk/83 ka/84 te/85 si/86 co/87 bu/88 vy/89

ag	ix	op	uk	ka	te	si	co	bu	vy
80	81	82	83	84	85	86	87	88	89

last 3 of the 5 letters:

B 80

1st	2nd	3rd	4th	5th	6th		7th	8th	9th	10th	11th	12th class
min	mio	mip	mir	mis	mit	50	miv	miw	mie	miz	mib	mic
moy	mop	mor	mos	mot	mov	51	mow	moe	moz	mob	moc	mod
mup	mur	mus	mut	muv	muw	52	mue	muz	mub	muc	mud	muf
myr	mys	myt	myv	myw	mye	53	myz	myb	myc	myd	myf	myg
nam	nan	nao	nap	nar	nas	54	nat	nav	naw	nae	naz	nab
nen	neo	nep	ner	nes	net	55	nev	ney	neu	nez	neb	nec
nio	nip	nir	nis	nit	niv	56	niw	nie	niz	nib	nic	nid
nop	nor	nos	not	nov	now	57	noe	noz	nob	noc	nod	nof
nur	nus	nut	nuv	nuw	nue	58	nuz	nub	nuc	nud	nuf	nug
nys	nyt	nyv	nyw	nye	nyz	59	nyb	nyc	nyd	nyf	nyg	nyh
pao	pap	par	pas	pat	pav	60	paw	pae	paz	pab	pac	pad
pep	per	pes	pet	pev	pey	61	peu	pez	peb	pec	ped	pef
pir	pis	pit	piv	piw	pie	62	piz	pib	pic	pid	pif	pig
pos	pot	pov	pow	poe	poz	63	pob	poc	pod	pof	pog	poh
put	puv	puw	pue	puz	pub	64	puc	pud	puf	pug	puh	pua
pyv	pyw	pye	pyz	pyb	pyc	65	pyd	pyf	pyg	pyh	pya	pyk
rap	rar	ras	rat	rav	raw	66	rae	raz	rab	rac	rad	raf
rer	res	ret	rev	rey	reu	67	rez	reb	rec	red	ref	reg
ris	rit	riv	riw	rie	riz	68	rib	ric	rid	rif	rig	rih
rot	rov	row	roe	roz	rob	69	roc	rod	rof	rog	roh	roa
ruv	ruw	rue	ruz	rub	ruc	70	rud	ruf	rug	ruh	rua	ruk
ryw	rye	ryz	ryb	ryc	ryd	71	ryf	ryg	ryh	rya	ryk	ryl
sar	sas	sat	sav	saw	sae	72	saz	sab	sac	sad	saf	sag
ses	set	sev	sey	seu	sez	73	seb	sec	sed	sef	seg	seh
sit	siv	siw	sie	siz	sib	74	sic	sid	sif	sig	sih	sia
sov	sow	soe	soz	sob	soc	75	sod	sof	sog	soh	soa	sok
suw	sue	suz	sub	suc	sud	76	suf	sug	suh	sua	suk	sul
sye	syz	syb	syc	syd	syf	77	syg	syh	sya	syk	syl	sym
tas	tat	tav	taw	tae	taz	78	tab	tac	tad	taf	tag	tah
tet	tev	tey	teu	tez	teb	79	tec	ted	tef	teg	teh	tea
tiv	tiw	tie	tiz	tib	tic	80	tid	tif	tig	tih	tia	tik
tow	toe	toz	tob	toc	tod	81	tof	tog	toh	toa	tok	tol
tue	tuz	tub	tuc	tud	tuf	82	tug	tuh	tua	tuk	tul	tum
tyz	tyb	tyc	tyd	tyf	tyg	83	tyh	tya	tyk	tyl	tym	tyn
vat	vav	vaw	vae	vaz	vab	84	vac	vad	vaf	vag	vah	vai
vev	vey	veu	vez	veb	vec	85	ved	vef	veg	veh	vea	vek
viw	vie	viz	vib	vic	vid	86	vif	vig	vih	via	vik	vil
voe	voz	vob	voc	vod	vof	87	vog	voh	voa	vok	vol	vom
vuz	vub	vuc	vud	vuf	vug	88	vuh	vua	vuk	vul	vum	vun
vyb	vyc	vyd	vyf	vyg	vyh	89	vya	vyk	vyl	vym	vyn	vyo
wav	waw	wae	waz	wab	wac	90	wad	waf	wag	wah	wai	wak
wey	weu	wez	web	wec	wed	91	wef	weg	weh	wea	wek	wel
wie	wiz	wib	wic	wid	wif	92	wig	wih	wia	wik	wil	wim
woz	wob	woc	wod	wof	wog	93	woh	woa	wok	wol	wom	won
zae	zaz	zab	zac	zad	zaf	94	zag	zah	zai	zak	zal	zam
zez	zeb	zec	zed	zef	zeg	95	zeh	zea	zek	zel	zem	zen
zib	zic	zid	zif	zig	zih	96	zia	zik	zil	zim	zin	zio
zoc	zod	zof	zog	zoh	zoa	97	zok	zol	zom	zon	zoy	zop
zud	zuf	zug	zuh	zua	zuk	98	zul	zum	zun	zuo	zup	zur
zyf	zyg	zyh	zya	zyk	zyl	99	zym	zyn	zyo	zyp	zyr	zys

B

First 2 of the 5 letters: er 90, ik 91, oc 92, ul 93, ha 94, je 95, ci 96, wo 97, du 98, zy 99

er	ik	oc	ul	ha	je	ci	wo	du	zy
90	91	92	93	94	95	96	97	98	99

last 3 of the 5 letters:

	1st	2nd	3rd	4th	5th	6th		7th	8th	9th	10th	11th	12th class
B	bab	bac	bad	baf	bag	bah	**00**	bai	bak	bal	bam	ban	bao
	bec	bed	bef	beg	beh	bea	**01**	bek	bel	bem	ben	beo	bep
90	bid	bif	big	bih	bia	bik	**02**	bil	bim	bin	bio	bip	bir
	bof	bog	boh	boa	bok	bol	**03**	bom	bon	boy	bop	bor	bos
	bug	buh	bua	buk	bul	bum	**04**	bun	buo	bup	bur	bus	but
	byh	bya	byk	byl	bym	byn	**05**	byo	byp	byr	bys	byt	byv
	cac	cad	caf	cag	cah	cai	**06**	cak	cal	cam	can	cao	cap
	ced	cef	ceg	ceh	cea	cek	**07**	cel	cem	cen	ceo	cep	cer
	cif	cig	cih	cia	cik	cil	**08**	cim	cin	cio	cip	cir	cis
	cog	coh	coa	cok	col	com	**09**	con	coy	cop	cor	cos	cot
	cuh	cua	cuk	cul	cum	cun	**10**	cuo	cup	cur	cus	cut	cuv
	cya	cyk	cyl	cym	cyn	cyo	**11**	cyp	cyr	cys	cyt	cyv	cyw
	dad	daf	dag	dah	dai	dak	**12**	dal	dam	dan	dao	dap	dar
	def	deg	deh	dea	dek	del	**13**	dem	den	deo	dep	der	des
	dig	dih	dia	dik	dil	dim	**14**	din	dio	dip	dir	dis	dit
	doh	doa	dok	dol	dom	don	**15**	doy	dop	dor	dos	dot	dov
	dua	duk	dul	dum	dun	duo	**16**	dup	dur	dus	dut	duv	duw
	dyk	dyl	dym	dyn	dyo	dyp	**17**	dyr	dys	dyt	dyv	dyw	dye
	faf	fag	fah	fai	fak	fal	**18**	fam	fan	fao	fap	far	fas
	feg	feh	fea	fek	fel	fem	**19**	fen	feo	fep	fer	fes	fet
	fih	fia	fik	fil	fim	fin	**20**	fio	fip	fir	fis	fit	fiv
	foa	fok	fol	fom	fon	foy	**21**	fop	for	fos	fot	fov	fow
	fuk	ful	fum	fun	fuo	fup	**22**	fur	fus	fut	fuv	fuw	fue
	fyl	fym	fyn	fyo	fyp	fyr	**23**	fys	fyt	fyv	fyw	fye	fyz
	gag	gah	gai	gak	gal	gam	**24**	gan	gao	gap	gar	gas	gat
	geh	gea	gek	gel	gem	gen	**25**	geo	gep	ger	ges	get	gev
	gia	gik	gil	gim	gin	gio	**26**	gip	gir	gis	git	giv	giw
	gok	gol	gom	gon	goy	gop	**27**	gor	gos	got	gov	gow	goe
	gul	gum	gun	guo	gup	gur	**28**	gus	gut	guv	guw	gue	guz
	gym	gyn	gyo	gyp	gyr	gys	**29**	gyt	gyv	gyw	gye	gyz	gyb
	hau	hai	hak	hal	ham	han	**30**	hao	hap	har	has	hat	hav
	hea	hek	hel	hem	hen	heo	**31**	hep	her	hes	het	hev	hey
	hik	hil	him	hin	hio	hip	**32**	hir	his	hit	hiv	hiw	hie
	hol	hom	hon	hoy	hop	hor	**33**	hos	hot	hov	how	hoe	hoz
	hum	hun	huo	hup	hur	hus	**34**	hut	huv	huw	hue	huz	hub
	hyn	hyo	hyp	hyr	hys	hyt	**35**	hyv	hyw	hye	hyz	hyb	hyc
	kak	kal	kam	kan	kao	kap	**36**	kar	kas	kat	kav	kaw	kae
	kel	kem	ken	keo	kep	ker	**37**	kes	ket	kev	key	keu	kez
	kim	kin	kio	kip	kir	kis	**38**	kit	kiv	kiw	kie	kiz	kib
	kon	koy	kop	kor	kos	kot	**39**	kov	kow	koe	koz	kob	koc
	kuo	kup	kur	kus	kut	kuv	**40**	kuw	kue	kuz	kub	kuc	kud
	kyp	kyr	kys	kyt	kyv	kyw	**41**	kye	kyz	kyb	kyc	kyd	kyf
	lal	lam	lan	lao	lap	lar	**42**	las	lat	lav	law	lae	laz
	lem	len	leo	lep	ler	les	**43**	let	lev	ley	leu	lez	leb
	lin	lio	lip	lir	lis	lit	**44**	liv	liw	lie	liz	lib	lic
	loy	lop	lor	los	lot	lov	**45**	low	loe	loz	lob	loc	lod
	lup	lur	lus	lut	luv	luw	**46**	lue	luz	lub	luc	lud	luf
	lyr	lys	lyt	lyv	lyw	lye	**47**	lyz	lyb	lyc	lyd	lyf	lyg
	mam	man	mao	map	mar	mas	**48**	mat	mav	maw	mae	maz	mab
	men	meo	mep	mer	mes	met	**49**	mev	mey	meu	mez	meb	mec

B

First 2 of the 5 letters	er	ik	oc	ul	ha	je	ci	wo	du	zy
	90	91	92	93	94	95	96	97	98	99

last 3 of the 5 letters:

1st	2nd	3rd	4th	5th	6th		7th	8th	9th	10th	11th	12th class
mio	mip	mir	mis	mit	miv	**50**	miw	mie	miz	mib	mic	mid
mop	mor	mos	mot	mov	mow	**51**	moe	moz	mob	moc	mod	mof
mur	mus	mut	muv	muw	mue	**52**	muz	mub	muc	mud	muf	mug
mys	myt	myv	myw	mye	myz	**53**	myb	myc	myd	myf	myg	myh
nan	nao	nap	nar	nas	nat	**54**	nav	naw	nae	naz	nab	nac
neo	nep	ner	nes	net	nev	**55**	ney	neu	nez	neb	nec	ned
nip	nir	nis	nit	niv	niw	**56**	nie	niz	nib	nic	nid	nif
nor	nos	not	nov	now	noe	**57**	noz	nob	noc	nod	nof	nog
nus	nut	nuv	nuw	nue	nuz	**58**	nub	nuc	nud	nuf	nug	nuh
nyt	nyv	nyw	nye	nyz	nyb	**59**	nyc	nyd	nyf	nyg	nyh	nya
pap	par	pas	pat	pav	paw	**60**	pae	paz	pab	pac	pad	paf
per	pes	pet	pev	pey	peu	**61**	pez	peb	pec	ped	pef	peg
pis	pit	piv	piw	pie	piz	**62**	pib	pic	pid	pif	pig	pih
pot	pov	pow	poe	poz	pob	**63**	poc	pod	pof	pog	poh	poa
puv	puw	pue	puz	pub	puc	**64**	pud	puf	pug	puh	pua	puk
pyw	pye	pyz	pyb	pyc	pyd	**65**	pyf	pyg	pyh	pya	pyk	pyl
rar	ras	rat	rav	raw	rae	**66**	raz	rab	rac	rad	raf	rag
res	ret	rev	rey	reu	rez	**67**	reb	rec	red	ref	reg	reh
rit	riv	riw	rie	riz	rib	**68**	ric	rid	rif	rig	rih	ria
rov	row	roe	roz	rob	roc	**69**	rod	rof	rog	roh	roa	rok
ruw	rue	ruz	rub	ruc	rud	**70**	ruf	rug	ruh	rua	ruk	rul
rye	ryz	ryb	ryc	ryd	ryf	**71**	ryg	ryh	rya	ryk	ryl	rym
sas	sat	sav	saw	sae	saz	**72**	sab	sac	sad	saf	sag	sah
set	sev	sey	seu	sez	seb	**73**	sec	sed	sef	seg	seh	sea
siv	siw	sie	siz	sib	sic	**74**	sid	sif	sig	sih	sia	sik
sow	soe	soz	sob	soc	sod	**75**	sof	sog	soh	soa	sok	sol
sue	suz	sub	suc	sud	suf	**76**	sug	suh	sua	suk	sul	sum
syz	syb	syc	syd	syf	syg	**77**	syh	sya	syk	syl	sym	syn
tat	tav	taw	tae	taz	tab	**78**	tac	tad	taf	tag	tah	tai
tev	tey	teu	tez	teb	tec	**79**	ted	tef	teg	teh	tea	tek
tiw	tie	tiz	tib	tic	tid	**80**	tif	tig	tih	tia	tik	til
toe	toz	tob	toc	tod	tof	**81**	tog	toh	toa	tok	tol	tom
tuz	tub	tuc	tud	tuf	tug	**82**	tuh	tua	tuk	tul	tum	tun
tyb	tyc	tyd	tyf	tyg	tyh	**83**	tya	tyk	tyl	tym	tyn	tyo
vav	vaw	vae	vaz	vab	vac	**84**	vad	vaf	vag	vah	vai	vak
vey	veu	vez	veb	vec	ved	**85**	vef	veg	veh	vea	vek	vel
vie	viz	vib	vic	vid	vif	**86**	vig	vih	via	vik	vil	vim
voz	vob	voc	vod	vof	vog	**87**	voh	voa	vok	vol	vom	von
vub	vuc	vud	vuf	vug	vuh	**88**	vua	vuk	vul	vum	vun	vuo
vyc	vyd	vyf	vyg	vyh	vya	**89**	vyk	vyl	vym	vyn	vyo	vyp
waw	wae	waz	wab	wac	wad	**90**	waf	wag	wah	wai	wak	wal
weu	wez	web	wec	wed	wef	**91**	weg	weh	wea	wek	wel	wem
wiz	wib	wic	wid	wif	wig	**92**	wih	wia	wik	wil	wim	win
wob	woc	wod	wof	wog	woh	**93**	woa	wok	wol	wom	won	woy
zaz	zab	zac	zad	zaf	zag	**94**	zah	zai	zak	zal	zam	zan
zeb	zec	zed	zef	zeg	zeh	**95**	zea	zek	zel	zem	zen	zeo
zic	zid	zif	zig	zih	zia	**96**	zik	zil	zim	zin	zio	zip
zod	zof	zog	zoh	zoa	zok	**97**	zol	zom	zon	zoy	zop	zor
zuf	zug	zuh	zua	zuk	zul	**98**	zum	zun	zuo	zup	zur	zus
zyg	zyh	zya	zyk	zyl	zym	**99**	zyn	zyo	zyp	zyr	zys	zyt

C

First 2 of the 5 letters:	ab	ip	on	um	ja	ke	ri	vo	hu	wy
last 3 of the 5 letters:	00	01	02	03	04	05	06	07	08	09

C 00

1st	2nd	3rd	4th	5th	6th		7th	8th	9th	10th	11th	12th class
aga	afa	ach	aca	ack	aci	**00**	abi	acu	abu	abo	acy	aby
ahk	ahl	ada	agd	agu	afu	**01**	adu	ado	aco	ady	aft	acs
aji	aku	agi	afi	aho	ago	**02**	afo	agy	afy	afs	ahi	adt
aln	alo	ahm	ahu	aja	ajy	**03**	ahy	ahs	ags	ake	aje	ahe
amo	any	aju	ajo	aky	alt	**04**	ajs	alw	ahr	alz	alu	ald
anf	apt	ako	aks	als	ams	**05**	akt	ame	ale	amp	and	amb
aps	arw	alp	aly	amt	arb	**06**	alv	anz	apd	arg	arn	ara
art	ass	amy	ant	ape	ase	**07**	ane	ard	arf	asu	asc	ati
asy	ate	arl	apf	arz	ath	**08**	arc	awa	ast	ata	ava	awu
azu	awd	ave	are	avi	aza	**09**	axa	azi	axi	awi	azo	axo
efa	ech	eca	eck	eci	ebi	**10**	ecu	ebu	ebo	ecy	eby	ebs
ehl	eda	egd	egu	efu	edu	**11**	edo	eco	edy	eft	ecs	edi
eku	egi	efi	eho	ego	efo	**12**	egy	efy	efs	ehi	edt	ege
elo	ehm	ehu	eja	ejy	ehy	**13**	ehs	egs	eke	eje	ehe	elc
eny	eju	ejo	eky	elt	ejs	**14**	elw	ehr	elz	elu	eld	eng
ept	eko	eks	els	ems	ekt	**15**	eme	ele	emp	end	emb	epa
erw	elp	ely	emt	erb	elv	**16**	enz	epd	erg	ern	era	erk
ess	emy	ent	epe	ese	ene	**17**	erd	erf	esu	esc	eti	eso
ete	erl	epf	erz	eth	erc	**18**	ewa	est	eta	eva	ewu	exu
ewd	eve	ere	evi	eza	exa	**19**	ezi	exi	ewi	ezo	exo	ezy
ich	ica	ick	ici	ibi	icu	**20**	ibu	ibo	icy	iby	ibs	ibt
ida	igd	igu	ifu	idu	ido	**21**	ico	idy	ift	ics	idi	ife
igi	ifi	iho	igo	ifo	igy	**22**	ify	ifs	ihi	idt	ige	ihb
ihm	ihu	ija	ijy	ihy	ihs	**23**	igs	ike	ije	ihe	ilc	ilf
iju	ijo	iky	ilt	ijs	ilw	**24**	ihr	ilz	ilu	ild	ing	ins
iko	iks	ils	ims	ikt	ime	**25**	ile	imp	ind	imb	ipa	iri
ilp	ily	imt	irb	ilv	inz	**26**	ipd	irg	irn	ira	irk	isa
imy	int	ipe	ise	ine	ird	**27**	irf	isu	isc	iti	iso	itu
irl	ipf	irz	ith	irc	iwa	**28**	ist	ita	iva	iwu	ixu	iwk
ive	ire	ivi	iza	ixa	izi	**29**	ixi	iwi	izo	ixo	izy	ixy
oca	ock	oci	obi	ocu	obu	**30**	obo	ocy	oby	obs	obt	ode
ogd	ogu	ofu	odu	odo	oco	**31**	ody	oft	ocs	odi	ofe	ogt
ofi	oho	ogo	ofo	ogy	ofy	**32**	ofs	ohi	odt	oge	ohb	ohp
ohu	oja	ojy	ohy	ohs	ogs	**33**	oke	oje	ohe	olc	olf	olg
ojo	oky	olt	ojs	olw	ohr	**34**	olz	olu	old	ong	ons	ona
oks	ols	oms	okt	ome	ole	**35**	omp	ond	omb	opa	ori	opi
oly	omt	orb	olv	onz	opd	**36**	org	orn	ora	ork	osa	orm
ont	ope	ose	one	ord	orf	**37**	osu	osc	oti	oso	otu	oto
opf	orz	oth	orc	owa	ost	**38**	ota	ova	owu	oxu	owk	ovu
ore	ovi	oza	oxa	ozi	oxi	**39**	owi	ozo	oxo	ozy	oxy	owy
uck	uci	ubi	ucu	ubu	ubo	**40**	ucy	uby	ubs	ubt	ude	uce
ugu	ufu	udu	udo	uco	udy	**41**	uft	ucs	udi	ufe	ugt	uhd
uho	ugo	ufo	ugy	ufy	ufs	**42**	uhi	udt	uge	uhb	uhp	ulb
uja	ujy	uhy	uhs	ugs	uke	**43**	uje	uhe	ulc	ulf	ulg	uma
uky	ult	ujs	ulw	uhr	ulz	**44**	ulu	uld	ung	uns	una	unk
uls	ums	ukt	ume	ule	ump	**45**	und	umb	upa	uri	upi	uru
umt	urb	ulv	unz	upd	urg	**46**	urn	ura	urk	usa	urm	usi
upe	use	une	urd	urf	usu	**47**	usc	uti	uso	utu	uto	uvo
urz	uth	urc	uwa	ust	uta	**48**	uva	uwu	uxu	uwk	uvu	uws
uvi	uza	uxa	uzi	uxi	uwi	**49**	uzo	uxo	uzy	uxy	uwy	uxt

C

First 2 of the 5 letters: { **az** 00 **ed** 01 **oj** 02 **us** 03 **va** 04 **ce** 05 **xi** 06 **do** 07 **tu** 08 **fy** 09

last 3 of the 5 letters:

1st	2nd	3rd	4th	5th	6th		7th	8th	9th	10th	11th	12th class
acy	aby	abs	abt	ade	ace	**50**	abe	agh	aba	act	aga	afa
aft	acs	adi	afe	agt	ahd	**51**	ads	aka	aht	aha	ahk	ahl
ahi	adt	age	ahb	ahp	alb	**52**	ahf	alk	ali	aki	aji	aku
aje	ahe	alc	alf	alg	ama	**53**	ala	ami	anu	alm	aln	alo
alu	ald	ang	ans	ana	ank	**54**	amd	anc	aph	amu	amo	any
and	amb	apa	ari	api	aru	**55**	ani	apo	ary	ano	anf	apt
arn	ara	ark	asa	arm	asi	**56**	apu	arp	atz	apy	aps	arw
asc	ati	aso	atu	ato	avo	**57**	aro	ats	avs	ars	art	ass
ava	awu	axu	awk	avu	aws	**58**	ask	avy	awo	asp	asy	ate
azo	axo	azy	axy	awy	axt	**59**	aty	aze	axe	awe	azu	awd
eby	ebs	ebt	ede	ece	ebe	**60**	egh	eba	ect	ega	efa	ech
ecs	edi	efe	egt	ehd	eds	**61**	eka	eht	eha	ehk	ehl	eda
edt	ege	ehb	ehp	elb	ehf	**62**	elk	eli	eki	eji	eku	egi
ehe	elc	elf	elg	ema	ela	**63**	emi	enu	elm	eln	elo	ehm
eld	eng	ens	ena	enk	emd	**64**	enc	eph	emu	emo	eny	eju
emb	epa	eri	epi	eru	eni	**65**	epo	ery	eno	enf	ept	eko
era	erk	esa	erm	esi	epu	**66**	erp	etz	epy	eps	erw	elp
eti	eso	etu	eto	evo	ero	**67**	ets	evs	ers	ert	ess	emy
ewu	exu	ewk	evu	ews	esk	**68**	evy	ewo	esp	esy	ete	erl
exo	ezy	exy	ewy	ext	ety	**69**	eze	exe	ewe	ezu	ewd	eve
ibs	ibt	ide	ice	ibe	igh	**70**	iba	ict	iga	ifa	ich	ica
idi	ife	igt	ihd	ids	ika	**71**	iht	iha	ihk	ihl	ida	igd
ige	ihb	ihp	ilb	ihf	ilk	**72**	ili	iki	iji	iku	igi	ifi
ilc	ilf	ilg	ima	ila	imi	**73**	inu	ilm	iln	ilo	ihm	ihu
ing	ins	ina	ink	imd	inc	**74**	iph	imu	imo	iny	iju	ijo
ipa	iri	ipi	iru	ini	ipo	**75**	iry	ino	inf	ipt	iko	iks
irk	isa	irm	isi	ipu	irp	**76**	itz	ipy	ips	irw	ilp	ily
iso	itu	ito	ivo	iro	its	**77**	ivs	irs	irt	iss	imy	int
ixu	iwk	ivu	iws	isk	ivy	**78**	iwo	isp	isy	ite	irl	ipf
izy	ixy	iwy	ixt	ity	ize	**79**	ixe	iwe	izu	iwd	ive	ire
obt	ode	oce	obe	ogh	oba	**80**	oct	oga	ofa	och	oca	ock
ofe	ogt	ohd	ods	oka	oht	**81**	oha	ohk	ohl	oda	ogd	ogu
ohb	ohp	olb	ohf	olk	oli	**82**	oki	oji	oku	ogi	ofi	oho
olf	olg	oma	ola	omi	onu	**83**	olm	oln	olo	ohm	ohu	oja
ons	ona	onk	omd	onc	oph	**84**	omu	omo	ony	oju	ojo	oky
ori	opi	oru	oni	opo	ory	**85**	ono	onf	opt	oko	oks	ols
osa	orm	osi	opu	orp	otz	**86**	opy	ops	orw	olp	oly	omt
otu	oto	ovo	oro	ots	ovs	**87**	ors	ort	oss	omy	ont	ope
owk	ovu	ows	osk	ovy	owo	**88**	osp	osy	ote	orl	opf	orz
oxy	owy	oxt	oty	oze	oxe	**89**	owe	ozu	owd	ove	ore	ovi
ude	uce	ube	ugh	uba	uct	**90**	uga	ufa	uch	uca	uck	uci
ugt	uhd	uds	uka	uht	uha	**91**	uhk	uhl	uda	ugd	ugu	ufu
uhp	ulb	uhf	ulk	uli	uki	**92**	uji	uku	ugi	ufi	uho	ugo
ulg	uma	ula	umi	unu	ulm	**93**	uln	ulo	uhm	uhu	uja	ujy
una	unk	umd	unc	uph	umu	**94**	umo	uny	uju	ujo	uky	ult
upi	uru	uni	upo	ury	uno	**95**	unf	upt	uko	uks	uls	ums
urm	usi	upu	urp	utz	upy	**96**	ups	urw	ulp	uly	umt	urb
uto	uvo	uro	uts	uvs	urs	**97**	urt	uss	umy	unt	upe	use
uvu	uws	usk	uvy	uwo	usp	**98**	usy	ute	url	upf	urz	uth
uwy	uxt	uty	uze	uxe	uwe	**99**	uzu	uwd	uve	ure	uvi	uza

C
00

C

First 2 of the 5 letters:	ad	em	iz	og	up	xa	ge	ni	bo	sy
last 3 of the 5 letters:	10	11	12	13	14	15	16	17	18	19

C 10

1st	2nd	3rd	4th	5th	6th		7th	8th	9th	10th	11th	12th class
afa	ach	aca	ack	aci	abi	**00**	acu	abu	abo	acy	aby	abs
ahl	ada	agd	agu	afu	adu	**01**	ado	aco	ady	aft	acs	adi
aku	agi	afi	aho	ago	afo	**02**	agy	afy	afs	ahi	adt	age
alo	ahm	ahu	aja	ajy	ahy	**03**	ahs	ags	ake	aje	ahe	alc
any	aju	ajo	aky	alt	ajs	**04**	alw	ahr	alz	alu	ald	ang
apt	ako	aks	als	ams	akt	**05**	ame	ale	amp	and	amb	apa
arw	alp	aly	amt	arb	alv	**06**	anz	apd	arg	arn	ara	ark
ass	amy	ant	ape	ase	ane	**07**	ard	arf	asu	asc	ati	aso
ate	arl	apf	arz	ath	arc	**08**	awa	ast	ata	ava	awu	axu
awd	ave	are	avi	aza	axa	**09**	azi	axi	awi	azo	axo	azy
ech	eca	eck	eci	ebi	ecu	**10**	ebu	ebo	ecy	eby	ebs	ebt
eda	egd	egu	efu	edu	edo	**11**	eco	edy	eft	ecs	edi	efe
egi	efi	eho	ego	efo	egy	**12**	efy	efs	ehi	edt	ege	ehb
ehm	ehu	eja	ejy	ehy	ehs	**13**	egs	eke	eje	ehe	elc	elf
eju	ejo	eky	elt	ejs	elw	**14**	ehr	elz	elu	eld	eng	ens
eko	eks	els	ems	ekt	eme	**15**	ele	emp	end	emb	epa	eri
elp	ely	emt	erb	elv	enz	**16**	epd	erg	ern	era	erk	esa
emy	ent	epe	ese	ene	erd	**17**	erf	esu	esc	eti	eso	etu
erl	epf	erz	eth	erc	ewa	**18**	est	eta	eva	ewu	exu	ewk
eve	ere	evi	eza	exa	ezi	**19**	exi	ewi	ezo	exo	ezy	exy
ica	ick	ici	ibi	icu	ibu	**20**	ibo	icy	iby	ibs	ibt	ide
igd	igu	ifu	idu	ido	ico	**21**	idy	ift	ics	idi	ife	igt
ifi	iho	igo	ifo	igy	ify	**22**	ifs	ihi	idt	ige	ihb	ihp
ihu	ija	ijy	ihy	ihs	igs	**23**	ike	ije	ihe	ilc	ilf	ilg
ijo	iky	ilt	ijs	ilw	ihr	**24**	ilz	ilu	ild	ing	ins	ina
iks	ils	ims	ikt	ime	ile	**25**	imp	ind	imb	ipa	iri	ipi
ily	imt	irb	ilv	inz	ipd	**26**	irg	irn	ira	irk	isa	irm
int	ipe	ise	ine	ird	irf	**27**	isu	isc	iti	iso	itu	ito
ipf	irz	ith	irc	iwa	ist	**28**	ita	iva	iwu	ixu	iwk	ivu
ire	ivi	iza	ixa	izi	ixi	**29**	iwi	izo	ixo	izy	ixy	iwy
ock	oci	obi	ocu	obu	obo	**30**	ocy	oby	obs	obt	ode	oce
ogu	ofu	odu	odo	oco	ody	**31**	oft	ocs	odi	ofe	ogt	ohd
oho	ogo	ofo	ogy	ofy	ofs	**32**	ohi	odt	oge	ohb	ohp	olb
oja	ojy	ohy	ohs	ogs	oke	**33**	oje	ohe	olc	olf	olg	oma
oky	olt	ojs	olw	ohr	olz	**34**	olu	old	ong	ons	ona	onk
ols	oms	okt	ome	ole	omp	**35**	ond	omb	opa	ori	opi	oru
omt	orb	olv	onz	opd	org	**36**	orn	ora	ork	osa	orm	osi
ope	ose	one	ord	orf	osu	**37**	osc	oti	oso	otu	oto	ovo
orz	oth	orc	owa	ost	ota	**38**	ova	owu	oxu	owk	ovu	ows
ovi	oza	oxa	ozi	oxi	owi	**39**	ozo	oxo	ozy	oxy	owy	oxt
uci	ubi	ucu	ubu	ubo	ucy	**40**	uby	ubs	ubt	ude	uce	ube
ufu	udu	udo	uco	udy	uft	**41**	ucs	udi	ufe	ugt	uhd	uds
ugo	ufo	ugy	ufy	ufs	uhi	**42**	udt	uge	uhb	uhp	ulb	uhf
ujy	uhy	uhs	ugs	uke	uje	**43**	uhe	ulc	ulf	ulg	uma	ula
ult	ujs	ulw	uhr	ulz	ulu	**44**	uld	ung	uns	una	unk	umd
ums	ukt	ume	ule	ump	und	**45**	umb	upa	uri	upi	uru	uni
urb	ulv	unz	upd	urg	urn	**46**	ura	urk	usa	urm	usi	upu
use	une	urd	urf	usu	usc	**47**	uti	uso	utu	uto	uvo	uro
uth	urc	uwa	ust	uta	uva	**48**	uwu	uxu	uwk	uvu	uws	usk
uza	uxa	uzi	uxi	uwi	uzo	**49**	uxo	uzy	uxy	uwy	uxt	uty

C

First 2 of the 5 letters { am 10 | ez 11 | iv 12 | ob 13 | la 14 | xe 15 | ti 16 | fo 17 | wu 18 | ky 19

last 3 of the 5 letters:

C
10

1st	2nd	3rd	4th	5th	6th		7th	8th	9th	10th	11th	12th class
aby	abs	abt	ade	ace	abe	50	agh	aba	act	aga	afa	ach
acs	adi	afe	agt	ahd	ads	51	aka	aht	aha	ahk	ahl	ada
adt	age	ahb	ahp	alb	ahf	52	alk	ali	aki	aji	aku	agi
ahe	alc	alf	alg	ama	ala	53	ami	anu	alm	aln	alo	ahm
ald	ang	ans	ana	ank	amd	54	anc	aph	amu	amo	any	aju
amb	apa	ari	api	aru	ani	55	apo	ary	ano	anf	apt	ako
ara	ark	asa	arm	asi	apu	56	arp	atz	apy	aps	arw	alp
ati	aso	atu	ato	avo	aro	57	ats	avs	ars	art	ass	amy
awu	axu	awk	avu	aws	ask	58	avy	awo	asp	asy	ate	arl
axo	azy	axy	awy	axt	aty	59	aze	axe	awe	azu	awd	ave
ebs	ebt	ede	ece	ebe	egh	60	eba	ect	ega	efa	ech	eca.
edi	efe	egt	ehd	eds	eka	61	eht	eha	ehk	ehl	eda	egd
ege	ehb	ehp	elb	ehf	elk	62	eli	eki	eji	eku	egi	efi
elc	elf	elg	ema	ela	emi	63	enu	elm	eln	elo	ehm	ehu
eng	ens	ena	enk	emd	enc	64	eph	emu	emo	eny	eju	ejo
epa	eri	epi	eru	eni	epo	65	ery	eno	enf	ept	eko	eks
erk	esa	erm	esi	epu	erp	66	etz	epy	eps	erw	elp	ely
eso	etu	eto	evo	ero	ets	67	evs	ers	ert	ess	emy	ent
exu	ewk	evu	ews	esk	evy	68	ewo	esp	esy	ete	erl	epf
ezy	exy	ewy	ext	ety	eze	69	exe	ewe	ezu	ewd	eve	ere
ibt	ide	ice	ibe	igh	iba	70	ict	iga	ifa	ich	ica	ick
ife	igt	ihd	ids	ika	iht	71	iha	ihk	ihl	ida	igd	igu
ihb	ihp	ilb	ihf	ilk	ili	72	iki	iji	iku	igi	ifi	iho
ilf	ilg	ima	ila	imi	inu	73	ilm	iln	ilo	ihm	ihu	ija
ins	ina	ink	imd	inc	iph	74	imu	imo	iny	iju	ijo	iky
iri	ipi	iru	ini	ipo	iry	75	ino	inf	ipt	iko	iks	ils
isa	irm	isi	ipu	irp	itz	76	ipy	ips	irw	ilp	ily	imt
itu	ito	ivo	iro	its	ivs	77	irs	irt	iss	imy	int	ipe
iwk	ivu	iws	isk	ivy	iwo	78	isp	isy	ite	irl	ipf	irz
ixy	iwy	ixt	ity	ize	ixe	79	iwe	izu	iwd	ive	ire	ivi
ode	oce	obe	ogh	oba	oct	80	oga	ofa	och	oca	ock	oci
ogt	ohd	ods	oka	oht	oha	81	ohk	ohl	oda	ogd	ogu	ofu
ohp	olb	ohf	olk	oli	oki	82	oji	oku	ogi	ofi	oho	ogo
olg	oma	ola	omi	onu	olm	83	oln	olo	ohm	ohu	oja	ojy
ona	onk	omd	onc	oph	omu	84	omo	ony	oju	ojo	oky	olt
opi	oru	oni	opo	ory	ono	85	onf	opt	oko	oks	ols	oms
orm	osi	opu	orp	otz	opy	86	ops	orw	olp	oly	omt	orb
oto	ovo	oro	ots	ovs	ors	87	ort	oss	omy	ont	ope	ose
ovu	ows	osk	ovy	owo	osp	88	osy	ote	orl	opf	orz	oth
owy	oxt	oty	oze	oxe	owe	89	ozu	owd	ove	ore	ovi	oza
uce	ube	ugh	uba	uct	uga	90	ufa	uch	uca	uck	uci	ubi
uhd	uds	uka	uht	uha	uhk	91	uhl	uda	ugd	ugu	ufu	udu
ulb	uhf	ulk	uli	uki	uji	92	uku	ugi	ufi	uho	ugo	ufo
uma	ula	umi	unu	ulm	uln	93	ulo	uhm	uhu	uja	ujy	uhy
unk	umd	unc	uph	umu	umo	94	uny	uju	ujo	uky	ult	ujs
uru	uni	upo	ury	uno	unf	95	upt	uko	uks	uls	ums	ukt
usi	upu	urp	utz	upy	ups	96	urw	ulp	uly	umt	urb	ulv
uvo	uro	uts	uvs	urs	urt	97	uss	umy	unt	upe	use	une
uws	usk	uvy	uwo	usp	usy	98	ute	url	upf	urz	uth	urc
uxt	uty	uze	uxe	uwe	uzu	99	uwd	uve	ure	uvi	uza	uxa

C

First 2 of the 5 letters:

af	ev	ib	ud	za	me	hi	to	cu	jy
20	21	22	23	24	25	26	27	28	29

last 3 of the 5 letters:

C
20

1st	2nd	3rd	4th	5th	6th		7th	8th	9th	10th	11th	12th class
ach	aca	ack	aci	abi	acu	**00**	abu	abo	acy	aby	abs	abt
ada	agd	agu	afu	adu	ado	**01**	aco	ady	aft	acs	adi	afe
agi	afi	aho	ago	afo	agy	**02**	afy	afs	ahi	adt	age	ahb
ahm	ahu	aja	ajy	ahy	ahs	**03**	ags	ake	aje	ahe	alc	alf
aju	ajo	aky	alt	ajs	alw	**04**	ahr	alz	alu	ald	ang	ans
ako	aks	als	ams	akt	ame	**05**	ale	amp	and	amb	apa	ari
alp	aly	amt	arb	alv	anz	**06**	apd	arg	arn	ara	ark	asa
amy	ant	ape	ase	ane	ard	**07**	arf	asu	asc	ati	aso	atu
arl	apf	arz	ath	arc	awa	**08**	ast	ata	ava	awu	axu	awk
ave	are	avi	aza	axa	azi	**09**	axi	awi	azo	axo	azy	axy
eca	eck	eci	ebi	ecu	ebu	**10**	ebo	ecy	eby	ebs	ebt	ede
egd	egu	efu	edu	edo	eco	**11**	edy	eft	ecs	edi	efe	egt
efi	eho	ego	efo	egy	efy	**12**	efs	ehi	edt	ege	ehb	ehp
ehu	eja	ejy	ehy	ehs	egs	**13**	eke	eje	ehe	elc	elf	elg
ejo	eky	elt	ejs	elw	ehr	**14**	elz	elu	eld	eng	ens	ena
eks	els	ems	ekt	eme	ele	**15**	emp	end	emb	epa	eri	epi
ely	emt	erb	elv	enz	epd	**16**	erg	ern	era	erk	esa	erm
ent	epe	ese	ene	erd	erf	**17**	esu	esc	eti	eso	etu	eto
epf	erz	eth	erc	ewa	est	**18**	eta	eva	ewu	exu	ewk	evu
ere	evi	eza	exa	ezi	exi	**19**	ewi	ezo	exo	ezy	exy	ewy
ick	ici	ibi	icu	ibu	ibo	**20**	icy	iby	ibs	ibt	ide	ice
igu	ifu	idu	ido	ico	idy	**21**	ift	ics	idi	ife	igt	ihd
iho	igo	ifo	igy	ify	ifs	**22**	ihi	idt	ige	ihb	ihp	ilb
ija	ijy	ihy	ihs	igs	ike	**23**	ije	ihe	ilc	ilf	ilg	ima
iky	ilt	ijs	ilw	ihr	ilz	**24**	ilu	ild	ing	ins	ina	ink
ils	ims	ikt	ime	ile	imp	**25**	ind	imb	ipa	iri	ipi	iru
imt	irb	ilv	inz	ipd	irg	**26**	irn	ira	irk	isa	irm	isi
ipe	ise	ine	ird	irf	isu	**27**	isc	iti	iso	itu	ito	ivo
irz	ith	irc	iwa	ist	ita	**28**	iva	iwu	ixu	iwk	ivu	iws
ivi	iza	ixa	izi	ixi	iwi	**29**	izo	ixo	izy	ixy	iwy	ixt
oci	obi	ocu	obu	obo	ocy	**30**	oby	obs	obt	ode	oce	obe
ofu	odu	odo	oco	ody	oft	**31**	ocs	odi	ofe	ogt	ohd	ods
ogo	ofo	ogy	ofy	ofs	ohi	**32**	odt	oge	ohb	ohp	olb	ohf
ojy	ohy	ohs	ogs	oke	oje	**33**	ohe	olc	olf	olg	oma	ola
olt	ojs	olw	ohr	olz	olu	**34**	old	ong	ons	ona	onk	omd
oms	okt	ome	ole	omp	ond	**35**	omb	opa	ori	opi	oru	oni
orb	olv	onz	opd	org	orn	**36**	ora	ork	osa	orm	osi	opu
ose	one	ord	orf	osu	osc	**37**	oti	oso	otu	oto	ovo	oro
oth	orc	owa	ost	ota	ova	**38**	owu	oxu	owk	ovu	ows	osk
oza	oxa	ozi	oxi	owi	ozo	**39**	oxo	ozy	oxy	owy	oxt	oty
ubi	ucu	ubu	ubo	ucy	uby	**40**	ubs	ubt	ude	uce	ube	ugh
udu	udo	uco	udy	uft	ucs	**41**	udi	ufe	ugt	uhd	uds	uka
ufo	ugy	ufy	ufs	uhi	udt	**42**	uge	uhb	uhp	ulb	uhf	ulk
uhy	uhs	ugs	uke	uje	uhe	**43**	ulc	ulf	ulg	uma	ula	umi
ujs	ulw	uhr	ulz	ulu	uld	**44**	ung	uns	una	unk	umd	unc
ukt	ume	ule	ump	und	umb	**45**	upa	uri	upi	uru	uni	upo
ulv	unz	upd	urg	urn	ura	**46**	urk	usa	urm	usi	upu	urp
une	urd	urf	usu	usc	uti	**47**	uso	utu	uto	uvo	uro	uts
urc	uwa	ust	uta	uva	uwu	**48**	uxu	uwk	uvu	uws	usk	uvy
uxa	uzi	uxi	uwi	uzo	uxo	**49**	uzy	uxy	uwy	uxt	uty	uze

C

First 2 of the 5 letters: ax 20 | ec 21 | ig 22 | ok 23 | uw 24 | ca 25 | di 26 | go 27 | lu 28 | ny 29

last 3 of the 5 letters:

1st	2nd	3rd	4th	5th	6th		7th	8th	9th	10th	11th	12th class
abs	abt	ade	ace	abe	agh	**50**	aba	act	aga	afa	ach	aca
adi	afe	agt	ahd	ads	aka	**51**	aht	aha	ahk	ahl	ada	agd
age	ahb	ahp	alb	ahf	alk	**52**	ali	aki	aji	aku	agi	afi
alc	alf	alg	ama	ala	ami	**53**	anu	alm	aln	alo	ahm	ahu
ang	ans	ana	ank	amd	anc	**54**	aph	amu	amo	any	aju	ajo
apa	ari	api	aru	ani	apo	**55**	ary	ano	anf	apt	ako	aks
ark	asa	arm	asi	apu	arp	**56**	atz	apy	aps	arw	alp	aly
aso	atu	ato	avo	aro	ats	**57**	avs	ars	art	ass	amy	ant
axu	awk	avu	aws	ask	avy	**58**	awo	asp	asy	ate	arl	apf
azy	axy	awy	axt	aty	aze	**59**	axe	awe	azu	awd	ave	are
ebt	ede	ece	ebe	egh	eba	**60**	ect	ega	efa	ech	eca	eck
efe	egt	ehd	eds	eka	eht	**61**	eha	ehk	ehl	eda	egd	egu
ehb	ehp	elb	ehf	elk	eli	**62**	eki	eji	eku	egi	efi	eho
elf	elg	ema	ela	emi	enu	**63**	elm	eln	elo	ehm	ehu	eja
ens	ena	enk	emd	enc	eph	**64**	emu	emo	eny	eju	ejo	eky
eri	epi	eru	eni	epo	ery	**65**	eno	enf	ept	eko	eks	els
esa	erm	esi	epu	erp	etz	**66**	epy	eps	erw	elp	ely	emt
etu	eto	evo	ero	ets	evs	**67**	ers	ert	ess	emy	ent	epe
ewk	evu	ews	esk	evy	ewo	**68**	esp	esy	ete	erl	epf	erz
exy	ewy	ext	ety	eze	exe	**69**	ewe	ezu	ewd	eve	ere	evi
ide	ice	ibe	igh	iba	ict	**70**	iga	ifa	ich	ica	ick	ici
igt	ihd	ids	ika	iht	iha	**71**	ihk	ihl	ida	igd	igu	ifu
ihp	ilb	ihf	ilk	ili	iki	**72**	iji	iku	igi	ifi	iho	igo
ilg	ima	ila	imi	inu	ilm	**73**	iln	ilo	ihm	ihu	ija	ijy
ina	ink	imd	inc	iph	imu	**74**	imo	iny	iju	ijo	iky	ilt
ipi	iru	ini	ipo	iry	ino	**75**	inf	ipt	iko	iks	ils	ims
irm	isi	ipu	irp	itz	ipy	**76**	ips	irw	ilp	ily	imt	irb
ito	ivo	iro	its	ivs	irs	**77**	irt	iss	imy	int	ipe	ise
ivu	iws	isk	ivy	iwo	isp	**78**	isy	ite	irl	ipf	irz	ith
iwy	ixt	ity	ize	ixe	iwe	**79**	izu	iwd	ive	ire	ivi	iza
oce	obe	ogh	oba	oct	oga	**80**	ofa	och	oca	ock	oci	obi
ohd	ods	oka	oht	oha	ohk	**81**	ohl	oda	ogd	ogu	ofu	odu
olb	ohf	olk	oli	oki	oji	**82**	oku	ogi	ofi	oho	ogo	ofo
oma	ola	omi	onu	olm	oln	**83**	olo	ohm	ohu	oja	ojy	ohy
onk	omd	onc	oph	omu	omo	**84**	ony	oju	ojo	oky	olt	ojs
oru	oni	opo	ory	ono	onf	**85**	opt	oko	oks	ols	oms	okt
osi	opu	orp	otz	opy	ops	**86**	orw	olp	oly	omt	orb	olv
ovo	oro	ots	ovs	ors	ort	**87**	oss	omy	ont	ope	ose	one
ows	osk	ovy	owo	osp	osy	**88**	ote	orl	opf	orz	oth	orc
oxt	oty	oze	oxe	owe	ozu	**89**	owd	ove	ore	ovi	oza	oxa
ube	ugh	uba	uct	uga	ufa	**90**	uch	uca	uck	uci	ubi	ucu
uds	uka	uht	uha	uhk	uhl	**91**	uda	ugd	ugu	ufu	udu	udo
uhf	ulk	uli	uki	uji	uku	**92**	ugi	ufi	uho	ugo	ufo	ugy
ula	umi	unu	ulm	uln	ulo	**93**	uhm	uhu	uja	ujy	uhy	uhs
umd	unc	uph	umu	umo	uny	**94**	uju	ujo	uky	ult	ujs	ulw
uni	upo	ury	uno	unf	upt	**95**	uko	uks	uls	ums	ukt	ume
upu	urp	utz	upy	ups	urw	**96**	ulp	uly	umt	urb	ulv	unz
uro	uts	uvs	urs	urt	uss	**97**	umy	unt	upe	use	une	urd
usk	uvy	uwo	usp	usy	ute	**98**	url	upf	urz	uth	urc	uwa
uty	uze	uxe	uwe	uzu	uwd	**99**	uve	ure	uvi	uza	uxa	uzi

C 20

C

First 2 of the 5 letters: { ac/30, et/31, ow/32, ux/33, ma/34, ve/35, li/36, ho/37, ru/38, xy/39

ac	et	ow	ux	ma	ve	li	ho	ru	xy
30	31	32	33	34	35	36	37	38	39

last 3 of the 5 letters:

C 30

1st	2nd	3rd	4th	5th	6th		7th	8th	9th	10th	11th	12th class
aca	ack	aci	abi	acu	abu	**00**	abo	acy	aby	abs	abt	ade
agd	agu	afu	adu	ado	aco	**01**	ady	aft	acs	adi	afe	agt
afi	aho	ago	afo	agy	afy	**02**	afs	ahi	adt	age	ahb	ahp
ahu	aja	ajy	ahy	ahs	ags	**03**	ake	aje	ahe	alc	alf	alg
ajo	aky	alt	ajs	alw	ahr	**04**	alz	alu	ald	ang	ans	ana
aks	als	ams	akt	ame	ale	**05**	amp	and	amb	apa	ari	api
aly	amt	arb	alv	anz	apd	**06**	arg	arn	ara	ark	asa	arm
ant	ape	ase	ane	ard	arf	**07**	asu	asc	ati	aso	atu	ato
apf	arz	ath	arc	awa	ast	**08**	ata	ava	awu	axu	awk	avu
are	avi	aza	axa	azi	axi	**09**	awi	azo	axo	azy	axy	awy
eck	eci	ebi	ecu	ebu	ebo	**10**	ecy	eby	ebs	ebt	ede	ece
egu	efu	edu	edo	eco	edy	**11**	eft	ecs	edi	efe	egt	ehd
eho	ego	efo	egy	efy	efs	**12**	ehi	edt	ege	ehb	ehp	elb
eja	ejy	ehy	ehs	egs	eke	**13**	eje	ehe	elc	elf	elg	ema
eky	elt	ejs	elw	ehr	elz	**14**	elu	eld	eng	ens	ena	enk
els	ems	ekt	eme	ele	emp	**15**	end	emb	epa	eri	epi	eru
emt	erb	elv	enz	epd	erg	**16**	ern	era	erk	esa	erm	esi
epe	ese	ene	erd	erf	esu	**17**	esc	eti	eso	etu	eto	evo
erz	eth	erc	ewa	est	eta	**18**	eva	ewu	exu	ewk	evu	ews
evi	eza	exa	ezi	exi	ewi	**19**	ezo	exo	ezy	exy	ewy	ext
ici	ibi	icu	ibu	ibo	icy	**20**	iby	ibs	ibt	ide	ice	ibe
ifu	idu	ido	ico	idy	ift	**21**	ics	idi	ife	igt	ihd	ids
igo	ifo	igy	ify	ifs	ihi	**22**	idt	ige	ihb	ihp	ilb	ihf
ijy	ihy	ihs	igs	ike	ije	**23**	ihe	ilc	ilf	ilg	ima	ila
ilt	ijs	ilw	ihr	ilz	ilu	**24**	ild	ing	ins	ina	ink	imd
ims	ikt	ime	ile	imp	ind	**25**	imb	ipa	iri	ipi	iru	ini
irb	ilv	inz	ipd	irg	irn	**26**	ira	irk	isa	irm	isi	ipu
ise	ine	ird	irf	isu	isc	**27**	iti	iso	itu	ito	ivo	iro
ith	irc	iwa	ist	ita	iva	**28**	iwu	ixu	iwk	ivu	iws	isk
iza	ixa	izi	ixi	iwi	izo	**29**	ixo	izy	ixy	iwy	ixt	ity
obi	ocu	obu	obo	ocy	oby	**30**	obs	obt	ode	oce	obe	ogh
odu	odo	oco	ody	oft	ocs	**31**	odi	ofe	ogt	ohd	ods	oka
ofo	ogy	ofy	ofs	ohi	odt	**32**	oge	ohb	ohp	olb	ohf	olk
ohy	ohs	ogs	oke	oje	ohe	**33**	olc	olf	olg	oma	ola	omi
ojs	olw	ohr	olz	olu	old	**34**	ong	ons	ona	onk	omd	onc
okt	ome	ole	omp	ond	omb	**35**	opa	ori	opi	oru	oni	opo
olv	onz	opd	org	orn	ora	**36**	ork	osa	orm	osi	opu	orp
one	ord	orf	osu	osc	oti	**37**	oso	otu	oto	ovo	oro	ots
orc	owa	ost	ota	ova	owu	**38**	oxu	owk	ovu	ows	osk	ovy
oxa	ozi	oxi	owi	ozo	oxo	**39**	ozy	oxy	owy	oxt	oty	oze
ucu	ubu	ubo	ucy	uby	ubs	**40**	ubt	ude	uce	ube	ugh	uba
udo	uco	udy	uft	ucs	udi	**41**	ufe	ugt	uhd	uds	uka	uht
ugy	ufy	ufs	uhi	udt	uge	**42**	uhb	uhp	ulb	uhf	ulk	uli
uhs	ugs	uke	uje	uhe	ulc	**43**	ulf	ulg	uma	ula	umi	unu
ulw	uhr	ulz	ulu	uld	ung	**44**	uns	una	unk	umd	unc	uph
ume	ule	ump	und	umb	upa	**45**	uri	upi	uru	uni	upo	ury
unz	upd	urg	urn	ura	urk	**46**	usa	urm	usi	upu	urp	utz
urd	urf	usu	usc	uti	uso	**47**	utu	uto	uvo	uro	uts	uvs
uwa	ust	uta	uva	uwu	uxu	**48**	uwk	uvu	uws	usk	uvy	uwo
uzi	uxi	uwi	uzo	uxo	uzy	**49**	uxy	uwy	uxt	uty	uze	uxe

C

First 2 of the 5 letters	as	eg	if	ol	un	be	wi	zo	su	py
	30	31	32	33	34	35	36	37	38	39

last 3 of the 5 letters:

1st	2nd	3rd	4th	5th	6th		7th	8th	9th	10th	11th	12th class
abt	ade	ace	abe	agh	aba	50	act	aga	afa	ach	aca	ack
afe	agt	ahd	ads	aka	aht	51	aha	ahk	ahl	ada	agd	agu
ahb	ahp	alb	ahf	alk	ali	52	aki	aji	aku	agi	afi	aho
alf	alg	ama	ala	ami	anu	53	alm	aln	alo	ahm	ahu	aja
ans	ana	ank	amd	anc	aph	54	amu	amo	any	aju	ajo	aky
ari	api	aru	ani	apo	ary	55	ano	anf	apt	ako	aks	als
asa	arm	asi	apu	arp	atz	56	apy	aps	arw	alp	aly	amt
atu	ato	avo	aro	ats	avs	57	ars	art	ass	amy	ant	ape
awk	avu	aws	ask	avy	awo	58	asp	asy	ate	arl	apf	arz
axy	awy	axt	aty	aze	axe	59	awe	azu	awd	ave	are	avi
ede	ece	ebe	egh	eba	ect	60	ega	efa	ech	eca	eck	eci
egt	ehd	eds	eka	eht	eha	61	ehk	ehl	eda	egd	egu	efu
ehp	elb	ehf	elk	eli	eki	62	eji	eku	egi	efi	eho	ego
elg	ema	ela	emi	enu	elm	63	eln	elo	ehm	ehu	eja	ejy
ena	enk	emd	enc	eph	emu	64	emo	eny	eju	ejo	eky	elt
epi	eru	eni	epo	ery	eno	65	enf	ept	eko	eks	els	ems
erm	esi	epu	erp	etz	epy	66	eps	erw	elp	ely	emt	erb
eto	evo	ero	ets	evs	ers	67	ert	ess	emy	ent	epe	ese
evu	ews	esk	evy	ewo	esp	68	esy	ete	erl	epf	erz	eth
ewy	ext	ety	eze	exe	ewe	69	ezu	ewd	eve	ere	evi	eza
ice	ibe	igh	iba	ict	iga	70	ifa	ich	ica	ick	ici	ibi
ihd	ids	ika	iht	iha	ihk	71	ihl	ida	igd	igu	ifu	idu
ilb	ihf	ilk	ili	iki	iji	72	iku	igi	ifi	iho	igo	ifo
ima	ila	imi	inu	ilm	iln	73	ilo	ihm	ihu	ija	ijy	ihy
ink	imd	inc	iph	imu	imo	74	iny	iju	ijo	iky	ilt	ijs
iru	ini	ipo	iry	ino	inf	75	ipt	iko	iks	ils	ims	ikt
isi	ipu	irp	itz	ipy	ips	76	irw	ilp	ily	imt	irb	ilv
ivo	iro	its	ivs	irs	irt	77	iss	imy	int	ipe	ise	ine
iws	isk	ivy	iwo	isp	isy	78	ite	irl	ipf	irz	ith	irc
ixt	ity	ize	ixe	iwe	izu	79	iwd	ive	ire	ivi	iza	ixa
obe	ogh	oba	oct	oga	ofa	80	och	oca	ock	oci	obi	ocu
ods	oka	oht	oha	ohk	ohl	81	oda	ogd	ogu	ofu	odu	odo
ohf	olk	oli	oki	oji	oku	82	ogi	ofi	oho	ogo	ofo	ogy
ola	omi	onu	olm	oln	olo	83	ohm	ohu	oja	ojy	ohy	ohs
omd	onc	oph	omu	omo	ony	84	oju	ojo	oky	olt	ojs	olw
oni	opo	ory	ono	onf	opt	85	oko	oks	ols	oms	okt	ome
opu	orp	otz	opy	ops	orw	86	olp	oly	omt	orb	olv	onz
oro	ots	ovs	ors	ort	oss	87	omy	ont	ope	ose	one	ord
osk	ovy	owo	osp	osy	ote	88	orl	opf	orz	oth	orc	owa
oty	oze	oxe	owe	ozu	owd	89	ove	ore	ovi	oza	oxa	ozi
ugh	uba	uct	uga	ufa	uch	90	uca	uck	uci	ubi	ucu	ubu
uka	uht	uha	uhk	uhl	uda	91	ugd	ugu	ufu	udu	udo	uco
ulk	uli	uki	uji	uku	ugi	92	ufi	uho	ugo	ufo	ugy	ufy
umi	unu	ulm	uln	ulo	uhm	93	uhu	uja	ujy	uhy	uhs	ugs
unc	uph	umu	umo	uny	uju	94	ujo	uky	ult	ujs	ulw	uhr
upo	ury	uno	unf	upt	uko	95	uks	uls	ums	ukt	ume	ule
urp	utz	upy	ups	urw	ulp	96	uly	umt	urb	ulv	unz	upd
uts	uvs	urs	urt	uss	umy	97	unt	upe	use	une	urd	urf
uvy	uwo	usp	usy	ute	url	98	upf	urz	uth	urc	uwa	ust
uze	uxe	uwe	uzu	uwd	uve	99	ure	uvi	uza	uxa	uzi	uxi

C

First 2 of the 5 letters: ak 40, es 41, il 42, ov 43, ur 44, na 45, le 46, ji 47, so 48, pu 49

last 3 of the 5 letters:

C 40

1st	2nd	3rd	4th	5th	6th		7th	8th	9th	10th	11th	12th class
ack	aci	abi	acu	abu	abo	00	acy	aby	abs	abt	ade	ace
agu	afu	adu	ado	aco	ady	01	aft	acs	adi	afe	agt	ahd
aho	ago	afo	agy	afy	afs	02	ahi	adt	age	ahb	ahp	alb
aja	ajy	ahy	ahs	ags	ake	03	aje	ahe	alc	alf	alg	ama
aky	alt	ajs	alw	ahr	alz	04	alu	ald	ang	ans	ana	ank
als	ams	akt	ame	ale	amp	05	and	amb	apa	ari	api	aru
amt	arb	alv	anz	apd	arg	06	arn	ara	ark	asa	arm	asi
ape	ase	ane	ard	arf	asu	07	asc	ati	aso	atu	ato	avo
arz	ath	arc	awa	ast	ata	08	ava	awu	axu	awk	avu	aws
avi	aza	axa	azi	axi	awi	09	azo	axo	azy	axy	awy	axt
eci	ebi	ecu	ebu	ebo	ecy	10	eby	ebs	ebt	ede	ece	ebe
efu	edu	edo	eco	edy	eft	11	ecs	edi	efe	egt	ehd	eds
ego	efo	egy	efy	efs	ehi	12	edt	ege	ehb	ehp	elb	ehf
ejy	ehy	ehs	egs	eke	eje	13	ehe	elc	elf	elg	ema	ela
elt	ejs	elw	ehr	elz	elu	14	eld	eng	ens	ena	enk	emd
ems	ekt	eme	ele	emp	end	15	emb	epa	eri	epi	eru	eni
erb	elv	enz	epd	erg	ern	16	era	erk	esa	erm	esi	epu
ese	ene	erd	erf	esu	esc	17	eti	eso	etu	eto	evo	ero
eth	erc	ewa	est	eta	eva	18	ewu	exu	ewk	evu	ews	esk
eza	exa	ezi	exi	ewi	ezo	19	exo	ezy	exy	ewy	ext	ety
ibi	icu	ibu	ibo	icy	iby	20	ibs	ibt	ide	ice	ibe	igh
idu	ido	ico	idy	ift	ics	21	idi	ife	igt	ihd	ids	ika
ifo	igy	ify	ifs	ihi	idt	22	ige	ihb	ihp	ilb	ihf	ilk
ihy	ihs	igs	ike	ije	ihe	23	ilc	ilf	ilg	ima	ila	imi
ijs	ilw	ihr	ilz	ilu	ild	24	ing	ins	ina	ink	imd	inc
ikt	ime	ile	imp	ind	imb	25	ipa	iri	ipi	iru	ini	ipo
ilv	inz	ipd	irg	irn	ira	26	irk	isa	irm	isi	ipu	irp
ine	ird	irf	isu	isc	iti	27	iso	itu	ito	ivo	iro	its
irc	iwa	ist	ita	iva	iwu	28	ixu	iwk	ivu	iws	isk	ivy
ixa	izi	ixi	iwi	izo	ixo	29	izy	ixy	iwy	ixt	ity	ize
ocu	obu	obo	ocy	oby	obs	30	obt	ode	oce	obe	ogh	oba
odo	oco	ody	oft	ocs	odi	31	ofe	ogt	ohd	ods	oka	oht
ogy	ofy	ofs	ohi	odt	oge	32	ohb	ohp	olb	ohf	olk	oli
ohs	ogs	oke	oje	ohe	olc	33	olf	olg	oma	ola	omi	onu
olw	ohr	olz	olu	old	ong	34	ons	ona	onk	omd	onc	oph
ome	ole	omp	ond	omb	opa	35	ori	opi	oru	oni	opo	ory
onz	opd	org	orn	ora	ork	36	osa	orm	osi	opu	orp	otz
ord	orf	osu	osc	oti	oso	37	otu	oto	ovo	oro	ots	ovs
owa	ost	ota	ova	owu	oxu	38	owk	ovu	ows	osk	ovy	owo
ozi	oxi	owi	ozo	oxo	ozy	39	oxy	owy	oxt	oty	oze	oxe
ubu	ubo	ucy	uby	ubs	ubt	40	ude	uce	ube	ugh	uba	uct
uco	udy	uft	ucs	udi	ufe	41	ugt	uhd	uds	uka	uht	uha
ufy	ufs	uhi	udt	uge	uhb	42	uhp	ulb	uhf	ulk	uli	uki
ugs	uke	uje	uhe	ulc	ulf	43	ulg	uma	ula	umi	unu	ulm
uhr	ulz	ulu	uld	ung	uns	44	una	unk	umd	unc	uph	umu
ule	ump	und	umb	upa	uri	45	upi	uru	uni	upo	ury	uno
upd	urg	urn	ura	urk	usa	46	urm	usi	upu	urp	utz	upy
urf	usu	usc	uti	uso	utu	47	uto	uvo	uro	uts	uvs	urs
ust	uta	uva	uwu	uxu	uwk	48	uvu	uws	usk	uvy	uwo	usp
uxi	uwi	uzo	uxo	uzy	uxy	49	uwy	uxt	uty	uze	uxe	uwe

C

First 2 of the 5 letters: ar 40, ef 41, ic 42, os 43, ga 44, de 45, pi 46, mo 47, nu 48, cy 49

First 2 of the 5 letters:	ar	ef	ic	os	ga	de	pi	mo	nu	cy
last 3 of the 5 letters:	40	41	42	43	44	45	46	47	48	49

C 40

1st	2nd	3rd	4th	5th	6th		7th	8th	9th	10th	11th	12th class
ade	ace	abe	agh	aba	act	**50**	aga	afa	ach	aca	ack	aci
agt	ahd	ads	aka	aht	aha	**51**	ahk	ahl	ada	agd	agu	afu
ahp	alb	ahf	alk	ali	aki	**52**	aji	aku	agi	afi	aho	ago
alg	ama	ala	ami	anu	alm	**53**	aln	alo	ahm	ahu	aja	ajy
ana	ank	amd	anc	aph	amu	**54**	amo	any	aju	ajo	aky	alt
api	aru	ani	apo	ary	ano	**55**	anf	apt	ako	aks	als	ams
arm	asi	apu	arp	atz	apy	**56**	aps	arw	alp	aly	amt	arb
ato	avo	aro	ats	avs	ars	**57**	art	ass	amy	ant	ape	ase
avu	aws	ask	avy	awo	asp	**58**	asy	ate	arl	apf	arz	ath
awy	axt	aty	aze	axe	awe	**59**	azu	awd	ave	are	avi	aza
ece	ebe	egh	eba	ect	ega	**60**	efa	ech	eca	eck	eci	ebi
ehd	eds	eka	eht	eha	ehk	**61**	ehl	eda	egd	egu	efu	edu
elb	ehf	elk	eli	eki	eji	**62**	eku	egi	efi	eho	ego	efo
ema	ela	emi	enu	elm	eln	**63**	elo	ehm	ehu	eja	ejy	ehy
enk	emd	enc	eph	emu	emo	**64**	eny	eju	ejo	eky	elt	ejs
eru	eni	epo	ery	eno	enf	**65**	ept	eko	eks	els	ems	ekt
esi	epu	erp	etz	epy	eps	**66**	erw	elp	ely	emt	erb	elv
evo	ero	ets	evs	ers	ert	**67**	ess	emy	ent	epe	ese	ene
ews	esk	evy	ewo	esp	esy	**68**	ete	erl	epf	erz	eth	erc
ext	ety	eze	exe	ewe	ezu	**69**	ewd	eve	ere	evi	eza	exa
ibe	igh	iba	ict	iga	ifa	**70**	ich	ica	ick	ici	ibi	icu
ids	ika	iht	iha	ihk	ihl	**71**	ida	igd	igu	ifu	idu	ido
ihf	ilk	ili	iki	iji	iku	**72**	igi	ifi	iho	igo	ifo	igy
ila	imi	inu	ilm	iln	ilo	**73**	ihm	ihu	ija	ijy	ihy	ihs
imd	inc	iph	imu	imo	iny	**74**	iju	ijo	iky	ilt	ijs	ilw
ini	ipo	iry	ino	inf	ipt	**75**	iko	iks	ils	ims	ikt	ime
ipu	irp	itz	ipy	ips	irw	**76**	ilp	ily	imt	irb	ilv	inz
iro	its	ivs	irs	irt	iss	**77**	imy	int	ipe	ise	ine	ird
isk	ivy	iwo	isp	isy	ite	**78**	irl	ipf	irz	ith	irc	iwa
ity	ize	ixe	iwe	izu	iwd	**79**	ive	ire	ivi	iza	ixa	izi
ogh	oba	oct	oga	ofa	och	**80**	oca	ock	oci	obi	ocu	obu
oka	oht	oha	ohk	ohl	oda	**81**	ogd	ogu	ofu	odu	odo	oco
olk	oli	oki	oji	oku	ogi	**82**	ofi	oho	ogo	ofo	ogy	ofy
omi	onu	olm	oln	olo	ohm	**83**	ohu	oja	ojy	ohy	ohs	ogs
onc	oph	omu	omo	ony	oju	**84**	ojo	oky	olt	ojs	olw	ohr
opo	ory	ono	onf	opt	oko	**85**	oks	ols	oms	okt	ome	ole
orp	otz	opy	ops	orw	olp	**86**	oly	omt	orb	olv	onz	opd
ots	ovs	ors	ort	oss	omy	**87**	ont	ope	ose	one	ord	orf
ovy	owo	osp	osy	ote	orl	**88**	opf	orz	oth	orc	owa	ost
oze	oxe	owe	ozu	owd	ove	**89**	ore	ovi	oza	oxa	ozi	oxi
uba	uct	uga	ufa	uch	uca	**90**	uck	uci	ubi	ucu	ubu	ubo
uht	uha	uhk	uhl	uda	ugd	**91**	ugu	ufu	udu	udo	uco	udy
uli	uki	uji	uku	ugi	ufi	**92**	uho	ugo	ufo	ugy	ufy	ufs
unu	ulm	uln	ulo	uhm	uhu	**93**	uja	ujy	uhy	uhs	ugs	uke
uph	umu	umo	uny	uju	ujo	**94**	uky	ult	ujs	ulw	uhr	ulz
ury	uno	unf	upt	uko	uks	**95**	uls	ums	ukt	ume	ule	ump
utz	upy	ups	urw	ulp	uly	**96**	umt	urb	ulv	unz	upd	urg
uvs	urs	urt	uss	umy	unt	**97**	upe	use	une	urd	urf	usu
uwo	usp	usy	ute	url	upf	**98**	urz	uth	urc	uwa	ust	uta
uxe	uwe	uzu	uwd	uve	ure	**99**	uvi	uza	uxa	uzi	uxi	uwi

C

First 2 of the 5 letters: { ex 50 | id 51 | of 52 | uz 53 | pa 54 | ne 55 | ki 56 | jo 57 | fu 58 | gy 59

last 3 of the 5 letters:

ex	id	of	uz	pa	ne	ki	jo	fu	gy
50	51	52	53	54	55	56	57	58	59

C 50

1st	2nd	3rd	4th	5th	6th		7th	8th	9th	10th	11th	12th class
aci	abi	acu	abu	abo	acy	**00**	aby	abs	abt	ade	ace	abe
afu	adu	ado	aco	ady	aft	**01**	acs	adi	afe	agt	ahd	ads
ago	afo	agy	afy	afs	ahi	**02**	adt	age	ahb	ahp	alb	ahf
ajy	ahy	ahs	ags	ake	aje	**03**	ahe	alc	alf	alg	ama	ala
alt	ajs	alw	ahr	alz	alu	**04**	ald	ang	ans	ana	ank	amd
ams	akt	ame	ale	amp	and	**05**	amb	apa	ari	api	aru	ani
arb	alv	anz	apd	arg	arn	**06**	ara	ark	asa	arm	asi	apu
ase	ane	ard	arf	asu	asc	**07**	ati	aso	atu	ato	avo	aro
ath	arc	awa	ast	ata	ava	**08**	awu	axu	awk	avu	aws	ask
aza	axa	azi	axi	awi	azo	**09**	axo	azy	axy	awy	axt	aty
ebi	ecu	ebu	ebo	ecy	eby	**10**	ebs	ebt	ede	ece	ebe	egh
edu	edo	eco	edy	eft	ecs	**11**	edi	efe	egt	ehd	eds	eka
efo	egy	efy	efs	ehi	edt	**12**	ege	ehb	ehp	elb	ehf	elk
ehy	ehs	egs	eke	eje	ehe	**13**	elc	elf	elg	ema	ela	emi
ejs	elw	ehr	elz	elu	eld	**14**	eng	ens	ena	enk	emd	enc
ekt	eme	ele	emp	end	emb	**15**	epa	eri	epi	eru	eni	epo
elv	enz	epd	erg	ern	era	**16**	erk	esa	erm	esi	epu	erp
ene	erd	erf	esu	esc	eti	**17**	eso	etu	eto	evo	ero	ets
erc	ewa	est	eta	eva	ewu	**18**	exu	ewk	evu	ews	esk	evy
exa	ezi	exi	ewi	ezo	exo	**19**	ezy	exy	ewy	ext	ety	eze
icu	ibu	ibo	icy	iby	ibs	**20**	ibt	ide	ice	ibe	igh	iba
ido	ico	idy	ift	ics	idi	**21**	ife	igt	ihd	ids	ika	iht
igy	ify	ifs	ihi	idt	ige	**22**	ihb	ihp	ilb	ihf	ilk	ili
ihs	igs	ike	ije	ihe	ilc	**23**	ilf	ilg	ima	ila	imi	inu
ilw	ihr	ilz	ilu	ild	ing	**24**	ins	ina	ink	imd	inc	iph
ime	ile	imp	ind	imb	ipa	**25**	iri	ipi	iru	ini	ipo	iry
inz	ipd	irg	irn	ira	irk	**26**	isa	irm	isi	ipu	irp	itz
ird	irf	isu	isc	iti	iso	**27**	itu	ito	ivo	iro	its	ivs
iwa	ist	ita	iva	iwu	ixu	**28**	iwk	ivu	iws	isk	ivy	iwo
izi	ixi	iwi	izo	ixo	izy	**29**	ixy	iwy	ixt	ity	ize	ixe
obu	obo	ocy	oby	obs	obt	**30**	ode	oce	obe	ogh	oba	oct
oco	ody	oft	ocs	odi	ofe	**31**	ogt	ohd	ods	oka	oht	oha
ofy	ofs	ohi	odt	oge	ohb	**32**	ohp	olb	ohf	olk	oli	oki
ogs	oke	oje	ohe	olc	olf	**33**	olg	oma	ola	omi	onu	olm
ohr	olz	olu	old	ong	ons	**34**	ona	onk	omd	onc	oph	omu
ole	omp	ond	omb	opa	ori	**35**	opi	oru	oni	opo	ory	ono
opd	org	orn	ora	ork	osa	**36**	orm	osi	opu	orp	otz	opy
orf	osu	osc	oti	oso	otu	**37**	oto	ovo	oro	ots	ovs	ors
ost	ota	ova	owu	oxu	owk	**38**	ovu	ows	osk	ovy	owo	osp
oxi	owi	ozo	oxo	ozy	oxy	**39**	owy	oxt	oty	oze	oxe	owe
ubo	ucy	uby	ubs	ubt	ude	**40**	uce	ube	ugh	uba	uct	uga
udy	uft	ucs	udi	ufe	ugt	**41**	uhd	uds	uka	uht	uha	uhk
ufs	uhi	udt	uge	uhb	uhp	**42**	ulb	uhf	ulk	uli	uki	uji
uke	uje	uhe	ulc	ulf	ulg	**43**	uma	ula	umi	unu	ulm	uln
ulz	ulu	uld	ung	uns	una	**44**	unk	umd	unc	uph	umu	umo
ump	und	umb	upa	uri	upi	**45**	uru	uni	upo	ury	uno	unf
urg	urn	ura	urk	usa	urm	**46**	usi	upu	urp	utz	upy	ups
usu	usc	uti	uso	utu	uto	**47**	uvo	uro	uts	uvs	urs	urt
uta	uva	uwu	uxu	uwk	uvu	**48**	uws	usk	uvy	uwo	usp	usy
uwi	uzo	uxo	uzy	uxy	uwy	**49**	uxt	uty	uze	uxe	uwe	uzu

C

First 2 of the 5 letters:

at	en	iw	oz	ba	pe	vi	ko	mu	ly
50	51	52	53	54	55	56	57	58	59

last 3 of the 5 letters:

1st	2nd	3rd	4th	5th	6th		7th	8th	9th	10th	11th	12th class
ace	abe	agh	aba	act	aga	**50**	afa	ach	aca	ack	aci	abi
ahd	ads	aka	aht	aha	ahk	**51**	ahl	ada	agd	agu	afu	adu
alb	ahf	alk	ali	aki	aji	**52**	aku	agi	afi	aho	ago	afo
ama	ala	ami	anu	alm	aln	**53**	alo	ahm	ahu	aja	ajy	ahy
ank	amd	anc	aph	amu	amo	**54**	any	aju	ajo	aky	alt	ajs
aru	ani	apo	ary	ano	anf	**55**	apt	ako	aks	als	ams	akt
asi	apu	arp	atz	apy	aps	**56**	arw	alp	aly	amt	arb	alv
avo	aro	ats	avs	ars	art	**57**	ass	amy	ant	ape	ase	ane
aws	ask	avy	awo	asp	asy	**58**	ate	arl	apf	arz	ath	arc
axt	aty	aze	axe	awe	azu	**59**	awd	ave	are	avi	aza	axa
ebe	egh	eba	ect	ega	efa	**60**	ech	eca.	eck	eci	ebi	ecu
eds	eka	eht	eha	ehk	ehl	**61**	eda	egd	egu	efu	edu	edo
ehf	elk	eli	eki	eji	eku	**62**	egi	efi	eho	ego	efo	egy
ela	emi	enu	elm	eln	elo	**63**	ehm	ehu	eja	ejy	ehy	ehs
emd	enc	eph	emu	emo	eny	**64**	eju	ejo	eky	elt	ejs	elw
eni	epo	ery	eno	enf	ept	**65**	eko	eks	els	ems	ekt	eme
epu	erp	etz	epy	eps	erw	**66**	elp	ely	emt	erb	elv	enz
ero	ets	evs	ers	ert	ess	**67**	emy	ent	epe	ese	ene	erd
esk	evy	ewo	esp	esy	ete	**68**	erl	epf	erz	eth	erc	ewa
ety	eze	exe	ewe	ezu	ewd	**69**	eve	ere	evi	eza	exa	ezi
igh	iba	ict	iga	ifa	ich	**70**	ica	ick	ici	ibi	icu	ibu
ika	iht	iha	ihk	ihl	ida	**71**	igd	igu	ifu	idu	ido	ico
ilk	ili	iki	iji	iku	igi	**72**	ifi	iho	igo	ifo	igy	ify
imi	inu	ilm	iln	ilo	ihm	**73**	ihu	ija	ijy	ihy	ihs	igs
inc	iph	imu	imo	iny	iju	**74**	ijo	iky	ilt	ijs	ilw	ihr
ipo	iry	ino	inf	ipt	iko	**75**	iks	ils	ims	ikt	ime	ile
irp	itz	ipy	ips	irw	ilp	**76**	ily	imt	irb	ilv	inz	ipd
its	ivs	irs	irt	iss	imy	**77**	int	ipe	ise	ine	ird	irf
ivy	iwo	isp	isy	ite	irl	**78**	ipf	irz	ith	irc	iwa	ist
ize	ixe	iwe	izu	iwd	ive	**79**	ire	ivi	iza	ixa	izi	ixi
oba	oct	oga	ofa	och	oca	**80**	ock	oci	obi	ocu	obu	obo
oht	oha	ohk	ohl	oda	ogd	**81**	ogu	ofu	odu	odo	oco	ody
oli	oki	oji	oku	ogi	ofi	**82**	oho	ogo	ofo	ogy	ofy	ofs
onu	olm	oln	olo	ohm	ohu	**83**	oja	ojy	ohy	ohs	ogs	oke
oph	omu	omo	ony	oju	ojo	**84**	oky	olt	ojs	olw	ohr	olz
ory	ono	onf	opt	oko	oks	**85**	ols	oms	okt	ome	ole	omp
otz	opy	ops	orw	olp	oly	**86**	omt	orb	olv	onz	opd	org
ovs	ors	ort	oss	omy	ont	**87**	ope	ose	one	ord	orf	osu
owo	osp	osy	ote	orl	opf	**88**	orz	oth	orc	owa	ost	ota
oxe	owe	ozu	owd	ove	ore	**89**	ovi	oza	oxa	ozi	oxi	owi
uct	uga	ufa	uch	uca	uck	**90**	uci	ubi	ucu	ubu	ubo	ucy
uha	uhk	uhl	uda	ugd	ugu	**91**	ufu	udu	udo	uco	udy	uft
uki	uji	uku	ugi	ufi	uho	**92**	ugo	ufo	ugy	ufy	ufs	uhi
ulm	uln	ulo	uhm	uhu	uja	**93**	ujy	uhy	uhs	ugs	uke	uje
umu	umo	uny	uju	ujo	uky	**94**	ult	ujs	ulw	uhr	ulz	ulu
uno	unf	upt	uko	uks	uls	**95**	ums	ukt	ume	ule	ump	und
upy	ups	urw	ulp	uly	umt	**96**	urb	ulv	unz	upd	urg	urn
urs	urt	uss	umy	unt	upe	**97**	use	une	urd	urf	usu	usc
usp	usy	ute	url	upf	urz	**98**	uth	urc	uwa	ust	uta	uva
uwe	uzu	uwd	uve	ure	uvi	**99**	uza	uxa	uzi	uxi	uwi	uzo

C

First 2 of the 5 letters:

aj	it	ox	uf	ra	se	zi	lo	xu	by
60	61	62	63	64	65	66	67	68	69

last 3 of the 5 letters:

C 60

1st	2nd	3rd	4th	5th	6th		7th	8th	9th	10th	11th	12th class
abi	acu	abu	abo	acy	aby	**00**	abs	abt	ade	ace	abe	agh
adu	ado	aco	ady	aft	acs	**01**	adi	afe	agt	ahd	ads	aka
afo	agy	afy	afs	ahi	adt	**02**	age	ahb	ahp	alb	ahf	alk
ahy	ahs	ags	ake	aje	ahe	**03**	alc	alf	alg	ama	ala	ami
ajs	alw	ahr	alz	alu	ald	**04**	ang	ans	ana	ank	amd	anc
akt	ame	ale	amp	and	amb	**05**	apa	ari	api	aru	ani	apo
alv	anz	apd	arg	arn	ara	**06**	ark	asa	arm	asi	apu	arp
ane	ard	arf	asu	asc	ati	**07**	aso	atu	ato	avo	aro	ats
arc	awa	ast	ata	ava	awu	**08**	axu	awk	avu	aws	ask	avy
axa	azi	axi	awi	azo	axo	**09**	azy	axy	awy	axt	aty	aze
ecu	ebu	ebo	ecy	eby	ebs	**10**	ebt	ede	ece	ebe	egh	eba
edo	eco	edy	eft	ecs	edi	**11**	efe	egt	ehd	eds	eka	eht
egy	efy	efs	ehi	edt	ege	**12**	ehb	ehp	elb	ehf	elk	eli
ehs	egs	eke	eje	ehe	elc	**13**	elf	elg	ema	ela	emi	enu
elw	ehr	elz	elu	eld	eng	**14**	ens	ena	enk	emd	enc	eph
eme	ele	emp	end	emb	epa	**15**	eri	epi	eru	eni	epo	ery
enz	epd	erg	ern	era	erk	**16**	esa	erm	esi	epu	erp	etz
erd	erf	esu	esc	eti	eso	**17**	etu	eto	evo	ero	ets	evs
ewa	est	eta	eva	ewu	exu	**18**	ewk	evu	ews	esk	evy	ewo
ezi	exi	ewi	ezo	exo	ezy	**19**	exy	ewy	ext	ety	eze	exe
ibu	ibo	icy	iby	ibs	ibt	**20**	ide	ice	ibe	igh	iba	ict
ico	idy	ift	ics	idi	ife	**21**	igt	ihd	ids	ika	iht	iha
ify	ifs	ihi	idt	ige	ihb	**22**	ihp	ilb	ihf	ilk	ili	iki
igs	ike	ije	ihe	ilc	ilf	**23**	ilg	ima	ila	imi	inu	ilm
ihr	ilz	ilu	ild	ing	ins	**24**	ina	ink	imd	inc	iph	imu
ile	imp	ind	imb	ipa	iri	**25**	ipi	iru	ini	ipo	iry	ino
ipd	irg	irn	ira	irk	isa	**26**	irm	isi	ipu	irp	itz	ipy
irf	isu	isc	iti	iso	itu	**27**	ito	ivo	iro	its	ivs	irs
ist	ita	iva	iwu	ixu	iwk	**28**	ivu	iws	isk	ivy	iwo	isp
ixi	iwi	izo	ixo	izy	ixy	**29**	iwy	ixt	ity	ize	ixe	iwe
obo	ocy	oby	obs	obt	ode	**30**	oce	obe	ogh	oba	oct	oga
ody	oft	ocs	odi	ofe	ogt	**31**	ohd	ods	oka	oht	oha	ohk
ofs	ohi	odt	oge	ohb	ohp	**32**	olb	ohf	olk	oli	oki	oji
oke	oje	ohe	olc	olf	olg	**33**	oma	ola	omi	onu	olm	oln
olz	olu	old	ong	ons	ona	**34**	onk	omd	onc	oph	omu	omo
omp	ond	omb	opa	ori	opi	**35**	oru	oni	opo	ory	ono	onf
org	orn	ora	ork	osa	orm	**36**	osi	opu	orp	otz	opy	ops
osu	osc	oti	oso	otu	oto	**37**	ovo	oro	ots	ovs	ors	ort
ota	ova	owu	oxu	owk	ovu	**38**	ows	osk	ovy	owo	osp	osy
owi	ozo	oxo	ozy	oxy	owy	**39**	oxt	oty	oze	oxe	owe	ozu
ucy	uby	ubs	ubt	ude	uce	**40**	ube	ugh	uba	uct	uga	ufa
uft	ucs	udi	ufe	ugt	uhd	**41**	uds	uka	uht	uha	uhk	uhl
uhi	udt	uge	uhb	uhp	ulb	**42**	uhf	ulk	uli	uki	uji	uku
uje	uhe	ulc	ulf	ulg	uma	**43**	ula	umi	unu	ulm	uln	ulo
ulu	uld	ung	uns	una	unk	**44**	umd	unc	uph	umu	umo	uny
und	umb	upa	uri	upi	uru	**45**	uni	upo	ury	uno	unf	upt
urn	ura	urk	usa	urm	usi	**46**	upu	urp	utz	upy	ups	urw
usc	uti	uso	utu	uto	uvo	**47**	uro	uts	uvs	urs	urt	uss
uva	uwu	uxu	uwk	uvu	uws	**48**	usk	uvy	uwo	usp	usy	ute
uzo	uxo	uzy	uxy	uwy	uxt	**49**	uty	uze	uxe	uwe	uzu	uwd

C

First 2 of the 5 letters: } aw 60, ek 61, is 62, om 63, ub 64, da 65, we 66, gi 67, no 68, zu 69

last 3 of the 5 letters:

1st	2nd	3rd	4th	5th	6th		7th	8th	9th	10th	11th	12th class
abe	agh	aba	act	aga	afa	**50**	ach	aca	ack	aci	abi	acu
ads	aka	aht	aha	ahk	ahl	**51**	ada	agd	agu	afu	adu	ado
ahf	alk	ali	aki	aji	aku	**52**	agi	afi	aho	ago	afo	agy
ala	ami	anu	alm	aln	alo	**53**	ahm	ahu	aja	ajy	ahy	ahs
amd	anc	aph	amu	amo	any	**54**	aju	ajo	aky	alt	ajs	alw
ani	apo	ary	ano	anf	apt	**55**	ako	aks	als	ams	akt	ame
apu	arp	atz	apy	aps	arw	**56**	alp	aly	amt	arb	alv	anz
aro	ats	avs	ars	art	ass	**57**	amy	ant	ape	ase	ane	ard
ask	avy	awo	asp	asy	ate	**58**	arl	apf	arz	ath	arc	awa
aty	aze	axe	awe	azu	awd	**59**	ave	are	avi	aza	axa	azi
egh	eba	ect	ega	efa	ech	**60**	eca	eck	eci	ebi	ecu	ebu
eka	eht	eha	ehk	ehl	eda	**61**	egd	egu	efu	edu	edo	eco
elk	eli	eki	eji	eku	egi	**62**	efi	eho	ego	efo	egy	efy
emi	enu	elm	eln	elo	ehm	**63**	ehu	eja	ejy	ehy	ehs	egs
enc	eph	emu	emo	eny	eju	**64**	ejo	eky	elt	ejs	elw	ehr
epo	ery	eno	enf	ept	eko	**65**	eks	els	ems	ekt	eme	ele
erp	etz	epy	eps	erw	elp	**66**	ely	emt	erb	elv	enz	epd
ets	evs	ers	ert	ess	emy	**67**	ent	epe	ese	ene	erd	erf
evy	ewo	esp	esy	ete	erl	**68**	epf	erz	eth	erc	ewa	est
eze	exe	ewe	ezu	ewd	eve	**69**	ere	evi	eza	exa	ezi	exi
iba	ict	iga	ifa	ich	ica	**70**	ick	ici	ibi	icu	ibu	ibo
iht	iha	ihk	ihl	ida	igd	**71**	igu	ifu	idu	ido	ico	idy
ili	iki	iji	iku	igi	ifi	**72**	iho	igo	ifo	igy	ify	ifs
inu	ilm	iln	ilo	ihm	ihu	**73**	ija	ijy	ihy	ihs	igs	ike
iph	imu	imo	iny	iju	ijo	**74**	iky	ilt	ijs	ilw	ihr	ilz
iry	ino	inf	ipt	iko	iks	**75**	ils	ims	ikt	ime	ile	imp
itz	ipy	ips	irw	ilp	ily	**76**	imt	irb	ilv	inz	ipd	irg
ivs	irs	irt	iss	imy	int	**77**	ipe	ise	ine	ird	irf	isu
iwo	isp	isy	ite	irl	ipf	**78**	irz	ith	irc	iwa	ist	ita
ixe	iwe	izu	iwd	ive	ire	**79**	ivi	iza	ixa	izi	ixi	iwi
oct	oga	ofa	och	oca	ock	**80**	oci	obi	ocu	obu	obo	ocy
oha	ohk	ohl	oda	ogd	ogu	**81**	ofu	odu	odo	oco	ody	oft
oki	oji	oku	ogi	ofi	oho	**82**	ogo	ofo	ogy	ofy	ofs	ohi
olm	oln	olo	ohm	ohu	oja	**83**	ojy	ohy	ohs	ogs	oke	oje
omu	omo	ony	oju	ojo	oky	**84**	olt	ojs	olw	ohr	olz	olu
ono	onf	opt	oko	oks	ols	**85**	oms	okt	ome	ole	omp	ond
opy	ops	orw	olp	oly	omt	**86**	orb	olv	onz	opd	org	orn
ors	ort	oss	omy	ont	ope	**87**	ose	one	ord	orf	osu	osc
osp	osy	ote	orl	opf	orz	**88**	oth	orc	owa	ost	ota	ova
owe	ozu	owd	ove	ore	ovi	**89**	oza	oxa	ozi	oxi	owi	ozo
uga	ufa	uch	uca	uck	uci	**90**	ubi	ucu	ubu	ubo	ucy	uby
uhk	uhl	uda	ugd	ugu	ufu	**91**	udu	udo	uco	udy	uft	ucs
uji	uku	ugi	ufi	uho	ugo	**92**	ufo	ugy	ufy	ufs	uhi	udt
uln	ulo	uhm	uhu	uja	ujy	**93**	uhy	uhs	ugs	uke	uje	uhe
umo	uny	uju	ujo	uky	ult	**94**	ujs	ulw	uhr	ulz	ulu	uld
unf	upt	uko	uks	uls	ums	**95**	ukt	ume	ule	ump	und	umb
ups	urw	ulp	uly	umt	urb	**96**	ulv	unz	upd	urg	urn	ura
urt	uss	umy	unt	upe	use	**97**	une	urd	urf	usu	usc	uti
usy	ute	url	upf	urz	uth	**98**	urc	uwa	ust	uta	uva	uwu
uzu	uwd	uve	ure	uvi	uza	**99**	uxa	uzi	uxi	uwi	uzo	uxo

C
60

C

First 2 of the 5 letters { al 70 | eb 71 | im 72 | ot 73 | ug 74 | sa 75 | fe 76 | bi 77 | ku 78 | ry 79

last 3 of the 5 letters:

C 70

1st	2nd	3rd	4th	5th	6th		7th	8th	9th	10th	11th	12th class
acu	abu	abo	acy	aby	abs	**00**	abt	ade	ace	abe	agh	aba
ado	aco	ady	aft	acs	adi	**01**	afe	agt	ahd	ads	aka	aht
agy	afy	afs	ahi	adt	age	**02**	ahb	ahp	alb	ahf	alk	ali
ahs	ags	ake	aje	ahe	alc	**03**	alf	alg	ama	ala	ami	anu
alw	ahr	alz	alu	ald	ang	**04**	ans	ana	ank	amd	anc	aph
ame	ale	amp	and	amb	apa	**05**	ari	api	aru	ani	apo	ary
anz	apd	arg	arn	ara	ark	**06**	asa	arm	asi	apu	arp	atz
ard	arf	asu	asc	ati	aso	**07**	atu	ato	avo	aro	ats	avs
awa	ast	ata	ava	awu	axu	**08**	awk	avu	aws	ask	avy	awo
azi	axi	awi	azo	axo	azy	**09**	axy	awy	axt	aty	aze	axe
ebu	ebo	ecy	eby	ebs	ebt	**10**	ede	ece	ebe	egh	eba	ect
eco	edy	eft	ecs	edi	efe	**11**	egt	ehd	eds	eka	eht	eha
efy	efs	ehi	edt	ege	ehb	**12**	ehp	elb	ehf	elk	eli	eki
egs	eke	eje	ehe	elc	elf	**13**	elg	ema	ela	emi	enu	elm
ehr	elz	elu	eld	eng	ens	**14**	ena	enk	emd	enc	eph	emu
ele	emp	end	emb	epa	eri	**15**	epi	eru	eni	epo	ery	eno
epd	erg	ern	era	erk	esa	**16**	erm	esi	epu	erp	etz	epy
erf	esu	esc	eti	eso	etu	**17**	eto	evo	ero	ets	evs	ers
est	eta	eva	ewu	exu	ewk	**18**	evu	ews	esk	evy	ewo	esp
exi	ewi	ezo	exo	ezy	exy	**19**	ewy	ext	ety	eze	exe	ewe
ibo	icy	iby	ibs	ibt	ide	**20**	ice	ibe	igh	iba	ict	iga
idy	ift	ics	idi	ife	igt	**21**	ihd	ids	ika	iht	iha	ihk
ifs	ihi	idt	ige	ihb	ihp	**22**	ilb	ihf	ilk	ili	iki	iji
ike	ije	ihe	ilc	ilf	ilg	**23**	ima	ila	imi	inu	ilm	iln
ilz	ilu	ild	ing	ins	ina	**24**	ink	imd	inc	iph	imu	imo
imp	ind	imb	ipa	iri	ipi	**25**	iru	ini	ipo	iry	ino	inf
irg	irn	ira	irk	isa	irm	**26**	isi	ipu	irp	itz	ipy	ips
isu	isc	iti	iso	itu	ito	**27**	ivo	iro	its	ivs	irs	irt
ita	iva	iwu	ixu	iwk	ivu	**28**	iws	isk	ivy	iwo	isp	isy
iwi	izo	ixo	izy	ixy	iwy	**29**	ixt	ity	ize	ixe	iwe	izu
ocy	oby	obs	obt	ode	oce	**30**	obe	ogh	oba	oct	oga	ofa
oft	ocs	odi	ofe	ogt	ohd	**31**	ods	oka	oht	oha	ohk	ohl
ohi	odt	oge	ohb	ohp	olb	**32**	ohf	olk	oli	oki	oji	oku
oje	ohe	olc	olf	olg	oma	**33**	ola	omi	onu	olm	oln	olo
olu	old	ong	ons	ona	onk	**34**	omd	onc	oph	omu	omo	ony
ond	omb	opa	ori	opi	oru	**35**	oni	opo	ory	ono	onf	opt
orn	ora	ork	osa	orm	osi	**36**	opu	orp	otz	opy	ops	orw
osc	oti	oso	otu	oto	ovo	**37**	oro	ots	ovs	ors	ort	oss
ova	owu	oxu	owk	ovu	ows	**38**	osk	ovy	owo	osp	osy	ote
ozo	oxo	ozy	oxy	owy	oxt	**39**	oty	oze	oxe	owe	ozu	owd
uby	ubs	ubt	ude	uce	ube	**40**	ugh	uba	uct	uga	ufa	uch
ucs	udi	ufe	ugt	uhd	uds	**41**	uka	uht	uha	uhk	uhl	uda
udt	uge	uhb	uhp	ulb	uhf	**42**	ulk	uli	uki	uji	uku	ugi
uhe	ulc	ulf	ulg	uma	ula	**43**	umi	unu	ulm	uln	ulo	uhm
uld	ung	uns	una	unk	umd	**44**	unc	uph	umu	umo	uny	uju
umb	upa	uri	upi	uru	uni	**45**	upo	ury	uno	unf	upt	uko
ura	urk	usa	urm	usi	upu	**46**	urp	utz	upy	ups	urw	ulp
uti	uso	utu	uto	uvo	uro	**47**	uts	uvs	urs	urt	uss	umy
uwu	uxu	uwk	uvu	uws	usk	**48**	uvy	uwo	usp	usy	ute	url
uxo	uzy	uxy	uwy	uxt	uty	**49**	uze	uxe	uwe	uzu	uwd	uve

C

First 2 of the 5 letters	av	ep	od	uc	fa	he	mi	ro	ju	ty
	70	71	72	73	74	75	76	77	78	79

last 3 of the 5 letters:

C 70

1st	2nd	3rd	4th	5th	6th		7th	8th	9th	10th	11th	12th class
agh	aba	act	aga	afa	ach	**50**	aca	ack	aci	abi	acu	abu
aka	aht	aha	ahk	ahl	ada	**51**	agd	agu	afu	adu	ado	aco
alk	ali	aki	aji	aku	agi	**52**	afi	aho	ago	afo	agy	afy
ami	anu	alm	aln	alo	ahm	**53**	ahu	aja	ajy	ahy	ahs	ags
anc	aph	amu	amo	any	aju	**54**	ajo	aky	alt	ajs	alw	ahr
apo	ary	ano	anf	apt	ako	**55**	aks	als	ams	akt	ame	ale
arp	atz	apy	aps	arw	alp	**56**	aly	amt	arb	alv	anz	apd
ats	avs	ars	art	ass	amy	**57**	ant	ape	ase	ane	ard	arf
avy	awo	asp	asy	ate	arl	**58**	apf	arz	ath	arc	awa	ast
aze	axe	awe	azu	awd	ave	**59**	are	avi	aza	axa	azi	axi
eba	ect	ega	efa	ech	eca.	**60**	eck	eci	ebi	ecu	ebu	ebo
eht	eha	ehk	ehl	eda	egd	**61**	egu	efu	edu	edo	eco	edy
eli	eki	eji	eku	egi	efi	**62**	eho	ego	efo	egy	efy	efs
enu	elm	eln	elo	ehm	ehu	**63**	eja	ejy	ehy	ehs	egs	eke
eph	emu	emo	eny	eju	ejo	**64**	eky	elt	ejs	elw	ehr	elz
ery	eno	enf	ept	eko	eks	**65**	els	ems	ekt	eme	ele	emp
etz	epy	eps	erw	elp	ely	**66**	emt	erb	elv	enz	epd	erg
evs	ers	ert	ess	emy	ent	**67**	epe	ese	ene	erd	erf	esu
ewo	esp	esy	ete	erl	epf	**68**	erz	eth	erc	ewa	est	eta
exe	ewe	ezu	ewd	eve	ere	**69**	evi	eza	exa	ezi	exi	ewi
ict	iga	ifa	ich	ica	ick	**70**	ici	ibi	icu	ibu	ibo	icy
iha	ihk	ihl	ida	igd	igu	**71**	ifu	idu	ido	ico	idy	ift
iki	iji	iku	igi	ifi	iho	**72**	igo	ifo	igy	ify	ifs	ihi
ilm	iln	ilo	ihm	ihu	ija	**73**	ijy	ihy	ihs	igs	ike	ije
imu	imo	iny	iju	ijo	iky	**74**	ilt	ijs	ilw	ihr	ilz	ilu
ino	inf	ipt	iko	iks	ils	**75**	ims	ikt	ime	ile	imp	ind
ipy	ips	irw	ilp	ily	imt	**76**	irb	ilv	inz	ipd	irg	irn
irs	irt	iss	imy	int	ipe	**77**	ise	ine	ird	irf	isu	isc
isp	isy	ite	irl	ipf	irz	**78**	ith	irc	iwa	ist	ita	iva
iwe	izu	iwd	ive	ire	ivi	**79**	iza	ixa	izi	ixi	iwi	izo
oga	ofa	och	oca	ock	oci	**80**	obi	ocu	obu	obo	ocy	oby
ohk	ohl	oda	ogd	ogu	ofu	**81**	odu	odo	oco	ody	oft	ocs
oji	oku	ogi	ofi	oho	ogo	**82**	ofo	ogy	ofy	ofs	ohi	odt
oln	olo	ohm	ohu	oja	ojy	**83**	ohy	ohs	ogs	oke	oje	ohe
omo	ony	oju	ojo	oky	olt	**84**	ojs	olw	ohr	olz	olu	old
onf	opt	oko	oks	ols	oms	**85**	okt	ome	ole	omp	ond	omb
ops	orw	olp	oly	omt	orb	**86**	olv	onz	opd	org	orn	ora
ort	oss	omy	ont	ope	ose	**87**	one	ord	orf	osu	osc	oti
osy	ote	orl	opf	orz	oth	**88**	orc	owa	ost	ota	ova	owu
ozu	owd	ove	ore	ovi	oza	**89**	oxa	ozi	oxi	owi	ozo	oxo
ufa	uch	uca	uck	uci	ubi	**90**	ucu	ubu	ubo	ucy	uby	ubs
uhl	uda	ugd	ugu	ufu	udu	**91**	udo	uco	udy	uft	ucs	udi
uku	ugi	ufi	uho	ugo	ufo	**92**	ugy	ufy	ufs	uhi	udt	uge
ulo	uhm	uhu	uja	ujy	uhy	**93**	uhs	ugs	uke	uje	uhe	ulc
uny	uju	ujo	uky	ult	ujs	**94**	ulw	uhr	ulz	ulu	uld	ung
upt	uko	uks	uls	ums	ukt	**95**	ume	ule	ump	und	umb	upa
urw	ulp	uly	umt	urb	ulv	**96**	unz	upd	urg	urn	ura	urk
uss	umy	unt	upe	use	une	**97**	urd	urf	usu	usc	uti	uso
ute	url	upf	urz	uth	urc	**98**	uwa	ust	uta	uva	uwu	uxu
uwd	uve	ure	uvi	uza	uxa	**99**	uzi	uxi	uwi	uzo	uxo	uzy

C

First 2 of the 5 letters:	ap	el	in	or	ut	ta	re	xo	vu	dy
	80	81	82	83	84	85	86	87	88	89

last 3 of the 5 letters:

C 80

1st	2nd	3rd	4th	5th	6th		7th	8th	9th	10th	11th	12th class
abu	abo	acy	aby	abs	abt	**00**	ade	ace	abe	agh	aba	act
aco	ady	aft	acs	adi	afe	**01**	agt	ahd	ads	aka	aht	aha
afy	afs	ahi	adt	age	ahb	**02**	ahp	alb	ahf	alk	ali	aki
ags	ake	aje	ahe	alc	alf	**03**	alg	ama	ala	ami	anu	alm
ahr	alz	alu	ald	ang	ans	**04**	ana	ank	amd	anc	aph	amu
ale	amp	and	amb	apa	ari	**05**	api	aru	ani	apo	ary	ano
apd	arg	arn	ara	ark	asa	**06**	arm	asi	apu	arp	atz	apy
arf	asu	asc	ati	aso	atu	**07**	ato	avo	aro	ats	avs	ars
ast	ata	ava	awu	axu	awk	**08**	avu	aws	ask	avy	awo	asp
axi	awi	azo	axo	azy	axy	**09**	awy	axt	aty	aze	axe	awe
ebo	ecy	eby	ebs	ebt	ede	**10**	ece	ebe	egh	eba	ect	ega
edy	eft	ecs	edi	efe	egt	**11**	ehd	eds	eka	eht	eha	ehk
efs	ehi	edt	ege	ehb	ehp	**12**	elb	ehf	elk	eli	eki	eji
eke	eje	ehe	elc	elf	elg	**13**	ema	ela	emi	enu	elm	eln
elz	elu	eld	eng	ens	ena	**14**	enk	emd	enc	eph	emu	emo
emp	end	emb	epa	eri	epi	**15**	eru	eni	epo	ery	eno	enf
erg	ern	era	erk	esa	erm	**16**	esi	epu	erp	etz	epy	eps
esu	esc	eti	eso	etu	eto	**17**	evo	ero	ets	evs	ers	ert
eta	eva	ewu	exu	ewk	evu	**18**	ews	esk	evy	ewo	esp	esy
ewi	ezo	exo	ezy	exy	ewy	**19**	ext	ety	eze	exe	ewe	ezu
icy	iby	ibs	ibt	ide	ice	**20**	ibe	igh	iba	ict	iga	ifa
ift	ics	idi	ife	igt	ihd	**21**	ids	ika	iht	iha	ihk	ihl
ihi	idt	ige	ihb	ihp	ilb	**22**	ihf	ilk	ili	iki	iji	iku
ije	ihe	ilc	ilf	ilg	ima	**23**	ila	imi	inu	ilm	iln	ilo
ilu	ild	ing	ins	ina	ink	**24**	imd	inc	iph	imu	imo	iny
ind	imb	ipa	iri	ipi	iru	**25**	ini	ipo	iry	ino	inf	ipt
irn	ira	irk	isa	irm	isi	**26**	ipu	irp	itz	ipy	ips	irw
isc	iti	iso	itu	ito	ivo	**27**	iro	its	ivs	irs	irt	iss
iva	iwu	ixu	iwk	ivu	iws	**28**	isk	ivy	iwo	isp	isy	ite
izo	ixo	izy	ixy	iwy	ixt	**29**	ity	ize	ixe	iwe	izu	iwd
oby	obs	obt	ode	oce	obe	**30**	ogh	oba	oct	oga	ofa	och
ocs	odi	ofe	ogt	ohd	ods	**31**	oka	oht	oha	ohk	ohl	oda
odt	oge	ohb	ohp	olb	ohf	**32**	olk	oli	oki	oji	oku	ogi
ohe	olc	olf	olg	oma	ola	**33**	omi	onu	olm	oln	olo	ohm
old	ong	ons	ona	onk	omd	**34**	onc	oph	omu	omo	ony	oju
omb	opa	ori	opi	oru	oni	**35**	opo	ory	ono	onf	opt	oko
ora	ork	osa	orm	osi	opu	**36**	orp	otz	opy	ops	orw	olp
oti	oso	otu	oto	ovo	oro	**37**	ots	ovs	ors	ort	oss	omy
owu	oxu	owk	ovu	ows	osk	**38**	ovy	owo	osp	osy	ote	orl
oxo	ozy	oxy	owy	oxt	oty	**39**	oze	oxe	owe	ozu	owd	ove
ubs	ubt	ude	uce	ube	ugh	**40**	uba	uct	uga	ufa	uch	uca
udi	ufe	ugt	uhd	uds	uka	**41**	uht	uha	uhk	uhl	uda	ugd
uge	uhb	uhp	ulb	uhf	ulk	**42**	uli	uki	uji	uku	ugi	ufi
ulc	ulf	ulg	uma	ula	umi	**43**	unu	ulm	uln	ulo	uhm	uhu
ung	uns	una	unk	umd	unc	**44**	uph	umu	umo	uny	uju	ujo
upa	uri	upi	uru	uni	upo	**45**	ury	uno	unf	upt	uko	uks
urk	usa	urm	usi	upu	urp	**46**	utz	upy	ups	urw	ulp	uly
uso	utu	uto	uvo	uro	uts	**47**	uvs	urs	urt	uss	umy	unt
uxu	uwk	uvu	uws	usk	uvy	**48**	uwo	usp	usy	ute	url	upf
uzy	uxy	uwy	uxt	uty	uze	**49**	uxe	uwe	uzu	uwd	uve	ure

C

First 2 of the 5 letters:	ag	ix	op	uk	ka	te	si	co	bu	vy
	80	81	82	83	84	85	86	87	88	89

last 3 of the 5 letters:

C
80

1st	2nd	3rd	4th	5th	6th		7th	8th	9th	10th	11th	12th class
aba	act	aga	afa	ach	aca	**50**	ack	aci	abi	acu	abu	abo
aht	aha	ahk	ahl	ada	agd	**51**	agu	afu	adu	ado	aco	ady
ali	aki	aji	aku	agi	afi	**52**	aho	ago	afo	agy	afy	afs
anu	alm	aln	alo	ahm	ahu	**53**	aja	ajy	ahy	ahs	ags	ake
aph	amu	amo	any	aju	ajo	**54**	aky	alt	ajs	alw	ahr	alz
ary	ano	anf	apt	ako	aks	**55**	als	ams	akt	ame	ale	amp
atz	apy	aps	arw	alp	aly	**56**	amt	arb	alv	anz	apd	arg
avs	ars	art	ass	amy	ant	**57**	ape	ase	ane	ard	arf	asu
awo	asp	asy	ate	arl	apf	**58**	arz	ath	arc	awa	ast	ata
axe	awe	azu	awd	ave	are	**59**	avi	aza	axa	azi	axi	awi
ect	**ega**	**efa**	**ech**	**eca**	**eck**	**60**	**eci**	**ebi**	**ecu**	**ebu**	**ebo**	**ecy**
eha	ehk	ehl	eda	egd	egu	**61**	efu	edu	edo	eco	edy	eft
eki	eji	eku	egi	efi	eho	**62**	ego	efo	egy	efy	efs	ehi
elm	eln	elo	ehm	ehu	eja	**63**	ejy	ehy	ehs	egs	eke	eje
emu	emo	eny	eju	ejo	eky	**64**	elt	ejs	elw	ehr	elz	elu
eno	enf	ept	eko	eks	els	**65**	ems	ekt	eme	ele	emp	end
epy	eps	erw	elp	ely	emt	**66**	erb	elv	enz	epd	erg	ern
ers	ert	ess	emy	ent	epe	**67**	ese	ene	erd	erf	esu	esc
esp	esy	ete	erl	epf	erz	**68**	eth	erc	ewa	est	eta	eva
ewe	ezu	ewd	eve	ere	evi	**69**	eza	exa	ezi	exi	ewi	ezo
iga	**ifa**	**ich**	**ica**	**ick**	**ici**	**70**	**ibi**	**icu**	**ibu**	**ibo**	**icy**	**iby**
ihk	ihl	ida	igd	igu	ifu	**71**	idu	ido	ico	idy	ift	ics
iji	iku	igi	ifi	iho	igo	**72**	ifo	igy	ify	ifs	ihi	idt
iln	ilo	ihm	ihu	ija	ijy	**73**	ihy	ihs	igs	ike	ije	ihe
imo	iny	iju	ijo	iky	ilt	**74**	ijs	ilw	ihr	ilz	ilu	ild
inf	ipt	iko	iks	ils	ims	**75**	ikt	ime	ile	imp	ind	imb
ips	irw	ilp	ily	imt	irb	**76**	ilv	inz	ipd	irg	irn	ira
irt	iss	imy	int	ipe	ise	**77**	ine	ird	irf	isu	isc	iti
isy	ite	irl	ipf	irz	ith	**78**	irc	iwa	ist	ita	iva	iwu
izu	iwd	ive	ire	ivi	iza	**79**	ixa	izi	ixi	iwi	izo	ixo
ofa	**och**	**oca**	**ock**	**oci**	**obi**	**80**	**ocu**	**obu**	**obo**	**ocy**	**oby**	**obs**
ohl	oda	ogd	ogu	ofu	odu	**81**	odo	oco	ody	oft	ocs	odi
oku	ogi	ofi	oho	ogo	ofo	**82**	ogy	ofy	ofs	ohi	odt	oge
olo	ohm	ohu	oja	ojy	ohy	**83**	ohs	ogs	oke	oje	ohe	olc
ony	oju	ojo	oky	olt	ojs	**84**	olw	ohr	olz	olu	old	ong
opt	oko	oks	ols	oms	okt	**85**	ome	ole	omp	ond	omb	opa
orw	olp	oly	omt	orb	olv	**86**	onz	opd	org	orn	ora	ork
oss	omy	ont	ope	ose	one	**87**	ord	orf	osu	osc	oti	oso
ote	orl	opf	orz	oth	orc	**88**	owa	ost	ota	ova	owu	oxu
owd	ove	ore	ovi	oza	oxa	**89**	ozi	oxi	owi	ozo	oxo	ozy
uch	**uca**	**uck**	**uci**	**ubi**	**ucu**	**90**	**ubu**	**ubo**	**ucy**	**uby**	**ubs**	**ubt**
uda	ugd	ugu	ufu	udu	udo	**91**	uco	udy	uft	ucs	udi	ufe
ugi	ufi	uho	ugo	ufo	ugy	**92**	ufy	ufs	uhi	udt	uge	uhb
uhm	uhu	uja	ujy	uhy	uhs	**93**	ugs	uke	uje	uhe	ulc	ulf
uju	ujo	uky	ult	ujs	ulw	**94**	uhr	ulz	ulu	uld	ung	uns
uko	uks	uls	ums	ukt	ume	**95**	ule	ump	und	umb	upa	uri
ulp	uly	umt	urb	ulv	unz	**96**	upd	urg	urn	ura	urk	usa
umy	unt	upe	use	une	urd	**97**	urf	usu	usc	uti	uso	utu
url	upf	urz	uth	urc	uwa	**98**	ust	uta	uva	uwu	uxu	uwk
uve	ure	uvi	uza	uxa	uzi	**99**	uxi	uwi	uzo	uxo	uzy	uxy

C

First 2 of the 5 letters: { an 90 | ej 91 | ir 92 | uv 93 | wa 94 | ze 95 | fi 96 | po 97 | gu 98 | my 99

last 3 of the 5 letters:

	1st	2nd	3rd	4th	5th	6th		7th	8th	9th	10th	11th	12th class
C	abo	acy	aby	abs	abt	ade	**00**	ace	abe	agh	aba	act	aga
	ady	aft	acs	adi	afe	agt	**01**	ahd	ads	aka	aht	aha	ahk
90	afs	ahi	adt	age	ahb	ahp	**02**	alb	ahf	alk	ali	aki	aji
	ake	aje	ahe	alc	alf	alg	**03**	ama	ala	ami	anu	alm	aln
	alz	alu	ald	ang	ans	ana	**04**	ank	amd	anc	aph	amu	amo
	amp	and	amb	apa	ari	api	**05**	aru	ani	apo	ary	ano	anf
	arg	arn	ara	ark	asa	arm	**06**	asi	apu	arp	atz	apy	aps
	asu	asc	ati	aso	atu	ato	**07**	avo	aro	ats	avs	ars	art
	ata	ava	awu	axu	awk	avu	**08**	aws	ask	avy	awo	asp	asy
	awi	azo	axo	azy	axy	awy	**09**	axt	aty	aze	axe	awe	azu
	ecy	eby	ebs	ebt	ede	ece	**10**	ebe	egh	eba	ect	ega	efa
	eft	ecs	edi	efe	egt	ehd	**11**	eds	eka	eht	eha	ehk	ehl
	ehi	edt	ege	ehb	ehp	elb	**12**	ehf	elk	eli	eki	eji	eku
	eje	ehe	elc	elf	elg	ema	**13**	ela	emi	enu	elm	eln	elo
	elu	eld	eng	ens	ena	enk	**14**	emd	enc	eph	emu	emo	eny
	end	emb	epa	eri	epi	eru	**15**	eni	epo	ery	eno	enf	ept
	ern	era	erk	esa	erm	esi	**16**	epu	erp	etz	epy	eps	erw
	esc	eti	eso	etu	eto	evo	**17**	ero	ets	evs	ers	ert	ess
	eva	ewu	exu	ewk	evu	ews	**18**	esk	evy	ewo	esp	esy	ete
	ezo	exo	ezy	exy	ewy	ext	**19**	ety	eze	exe	ewe	ezu	ewd
	iby	ibs	ibt	ide	ice	ibe	**20**	igh	iba	ict	iga	ifa	ich
	ics	idi	ife	igt	ihd	ids	**21**	ika	iht	iha	ihk	ihl	ida
	idt	ige	ihb	ihp	ilb	ihf	**22**	ilk	ili	iki	iji	iku	igi
	ihe	ilc	ilf	ilg	ima	ila	**23**	imi	inu	ilm	iln	ilo	ihm
	ild	ing	ins	ina	ink	imd	**24**	inc	iph	imu	imo	iny	iju
	imb	ipa	iri	ipi	iru	ini	**25**	ipo	iry	ino	inf	ipt	iko
	ira	irk	isa	irm	isi	ipu	**26**	irp	itz	ipy	ips	irw	ilp
	iti	iso	itu	ito	ivo	iro	**27**	its	ivs	irs	irt	iss	imy
	iwu	ixu	iwk	ivu	iws	isk	**28**	ivy	iwo	isp	isy	ite	irl
	ixo	izy	ixy	iwy	ixt	ity	**29**	ize	ixe	iwe	izu	iwd	ive
	obs	obt	ode	oce	obe	ogh	**30**	oba	oct	oga	ofa	och	oca
	odi	ofe	ogt	ohd	ods	oka	**31**	oht	oha	ohk	ohl	oda	ogd
	oge	ohb	ohp	olb	ohf	olk	**32**	oli	oki	oji	oku	ogi	ofi
	olc	olf	olg	oma	ola	omi	**33**	onu	olm	oln	olo	ohm	ohu
	ong	ons	ona	onk	omd	onc	**34**	oph	omu	omo	ony	oju	ojo
	opa	ori	opi	oru	oni	opo	**35**	ory	ono	onf	opt	oko	oks
	ork	osa	orm	osi	opu	orp	**36**	otz	opy	ops	orw	olp	oly
	oso	otu	oto	ovo	oro	ots	**37**	ovs	ors	ort	oss	omy	ont
	oxu	owk	ovu	ows	osk	ovy	**38**	owo	osp	osy	ote	orl	opf
	ozy	oxy	owy	oxt	oty	oze	**39**	oxe	owe	ozu	owd	ove	ore
	ubt	ude	uce	ube	ugh	uba	**40**	uct	uga	ufa	uch	uca	uck
	ufe	ugt	uhd	uds	uka	uht	**41**	uha	uhk	uhl	uda	ugd	ugu
	uhb	uhp	ulb	uhf	ulk	uli	**42**	uki	uji	uku	ugi	ufi	uho
	ulf	ulg	uma	ula	umi	unu	**43**	ulm	uln	ulo	uhm	uhu	uja
	uns	una	unk	umd	unc	uph	**44**	umu	umo	uny	uju	ujo	uky
	uri	upi	uru	uni	upo	ury	**45**	uno	unf	upt	uko	uks	uls
	usa	urm	usi	upu	urp	utz	**46**	upy	ups	urw	ulp	uly	umt
	utu	uto	uvo	uro	uts	uvs	**47**	urs	urt	uss	umy	unt	upe
	uwk	uvu	uws	usk	uvy	uwo	**48**	usp	usy	ute	url	upf	urz
	uxy	uwy	uxt	uty	uze	uxe	**49**	uwe	uzu	uwd	uve	ure	uvi

First 2 of the 5 letters: er 90, ik 91, oc 92, ul 93, ha 94, je 95, ci 96, wo 97, du 98, zy 99

last 3 of the 5 letters:

1st	2nd	3rd	4th	5th	6th		7th	8th	9th	10th	11th	12th class
act	aga	afa	ach	aca	ack	**50**	aci	abi	acu	abu	abo	acy
aha	ahk	ahl	ada	agd	agu	**51**	afu	adu	ado	aco	ady	aft
aki	aji	aku	agi	afi	aho	**52**	ago	afo	agy	afy	afs	ahi
alm	aln	alo	ahm	ahu	aja	**53**	ajy	ahy	ahs	ags	ake	aje
amu	amo	any	aju	ajo	aky	**54**	alt	ajs	alw	ahr	alz	alu
ano	anf	apt	ako	aks	als	**55**	ams	akt	ame	ale	amp	and
apy	aps	arw	alp	aly	amt	**56**	arb	alv	anz	apd	arg	arn
ars	art	ass	amy	ant	ape	**57**	ase	ane	ard	arf	asu	asc
asp	asy	ate	arl	apf	arz	**58**	ath	arc	awa	ast	ata	ava
awe	azu	awd	ave	are	avi	**59**	aza	axa	azi	axi	awi	azo
ega	efa	ech	eca.	eck	eci	**60**	ebi	ecu	ebu	ebo	ecy	eby
ehk	ehl	eda	egd	egu	efu	**61**	edu	edo	eco	edy	eft	ecs
eji	eku	egi	efi	eho	ego	**62**	efo	egy	efy	efs	ehi	edt
eln	elo	ehm	ehu	eja	ejy	**63**	ehy	ehs	egs	eke	eje	ehe
emo	eny	eju	ejo	eky	elt	**64**	ejs	elw	ehr	elz	elu	eld
enf	ept	eko	eks	els	ems	**65**	ekt	eme	ele	emp	end	emb
eps	erw	elp	ely	emt	erb	**66**	elv	enz	epd	erg	ern	era
ert	ess	emy	ent	epe	ese	**67**	ene	erd	erf	esu	esc	eti
esy	ete	erl	epf	erz	eth	**68**	erc	ewa	est	eta	eva	ewu
ezu	ewd	eve	ere	evi	eza	**69**	exa	ezi	exi	ewi	ezo	exo
ifa	ich	ica	ick	ici	ibi	**70**	icu	ibu	ibo	icy	iby	ibs
ihl	ida	igd	igu	ifu	idu	**71**	ido	ico	idy	ift	ics	idi
iku	igi	ifi	iho	igo	ifo	**72**	igy	ify	ifs	ihi	idt	ige
ilo	ihm	ihu	ija	ijy	ihy	**73**	ihs	igs	ike	ije	ihe	ilc
iny	iju	ijo	iky	ilt	ijs	**74**	ilw	ihr	ilz	ilu	ild	ing
ipt	iko	iks	ils	ims	ikt	**75**	ime	ile	imp	ind	imb	ipa
irw	ilp	ily	imt	irb	ilv	**76**	inz	ipd	irg	irn	ira	irk
iss	imy	int	ipe	ise	ine	**77**	ird	irf	isu	isc	iti	iso
ite	irl	ipf	irz	ith	irc	**78**	iwa	ist	ita	iva	iwu	ixu
iwd	ive	ire	ivi	iza	ixa	**79**	izi	ixi	iwi	izo	ixo	izy
och	oca	ock	oci	obi	ocu	**80**	obu	obo	ocy	oby	obs	obt
oda	ogd	ogu	ofu	odu	odo	**81**	oco	ody	oft	ocs	odi	ofe
ogi	ofi	oho	ogo	ofo	ogy	**82**	ofy	ofs	ohi	odt	oge	ohb
ohm	ohu	oja	ojy	ohy	ohs	**83**	ogs	oke	oje	ohe	olc	olf
oju	ojo	oky	olt	ojs	olw	**84**	ohr	olz	olu	old	ong	ons
oko	oks	ols	oms	okt	ome	**85**	ole	omp	ond	omb	opa	ori
olp	oly	omt	orb	olv	onz	**86**	opd	org	orn	ora	ork	osa
omy	ont	ope	ose	one	ord	**87**	orf	osu	osc	oti	oso	otu
orl	opf	orz	oth	orc	owa	**88**	ost	ota	ova	owu	oxu	owk
ove	ore	ovi	oza	oxa	ozi	**89**	oxi	owi	ozo	oxo	ozy	oxy
uca	uck	uci	ubi	ucu	ubu	**90**	ubo	ucy	uby	ubs	ubt	ude
ugd	ugu	ufu	udu	udo	uco	**91**	udy	uft	ucs	udi	ufe	ugt
ufi	uho	ugo	ufo	ugy	ufy	**92**	ufs	uhi	udt	uge	uhb	uhp
uhu	uja	ujy	uhy	uhs	ugs	**93**	uke	uje	uhe	ulc	ulf	ulg
ujo	uky	ult	ujs	ulw	uhr	**94**	ulz	ulu	uld	ung	uns	una
uks	uls	ums	ukt	ume	ule	**95**	ump	und	umb	upa	uri	upi
uly	umt	urb	ulv	unz	upd	**96**	urg	urn	ura	urk	usa	urm
unt	upe	use	une	urd	urf	**97**	usu	usc	uti	uso	utu	uto
upf	urz	uth	urc	uwa	ust	**98**	uta	uva	uwu	uxu	uwk	uvu
ure	uvi	uza	uxa	uzi	uxi	**99**	uwi	uzo	uxo	uzy	uxy	uwy

Bei Fragen zur Produktsicherheit wenden Sie sich bitte an:
If you have any questions regarding product safety,
please contact:

Walter de Gruyter GmbH
Genthiner Straße 13
10785 Berlin
productsafety@degruyterbrill.com